DICTIONARY OF ELECTRONICS

DICTIONARY OF ELECTRONICS

V.K. PURI
M.Sc., Ph.D.

P. CHAND
M.Sc., Ph.D.

ANMOL PUBLICATIONS PVT. LTD.
NEW DELHI - 110 002 (INDIA)

ANMOL PUBLICATIONS PVT. LTD.
H.O.: 4374/4B, Ansari Road, Darya Ganj,
New Delhi-110 002 (India)
Ph.: 23278000, 23261597

B.O.: No. 1015, Ist Main Road, BSK IIIrd Stage
IIIrd Phase, IIIrd Block,
Bangalore - 560 085 (India)
Visit us at: www.anmolpublications.com

Dictionary of Electronics

First Edition, 1987

Second Revised Edition, 1994

Reprint 1995, 1996, 1998, 2000, 2004

Reprint, 2006

PRINTED IN INDIA

Printed at Mehra Offset Press, Delhi.

Preface

As the applications of electronics infiltrate our society and more people get intrigued, involved and confused by new terms and new meanings for old terms, defintions have to constantly updated and refined. This dictionary included the most frequently used words and terms involved in the electronics. A careful attempt has been made to select the terms because these are most likely to comfort the movice when dealing with this emerging technology. Further, emphasis has also been laid on the latest developments in the applications of electronics.

This dictionary will be of value to students of electronics, physicists, engineering students. This dictionary is a correlated condensation of electronic concepts designed for time saving references work.

An attempt has been made to write the entries in a clear and lucid style to provide both straight forward definitions and invaluable background information. Line drawings have been included whenever the meaning of a word can be best understood by means of a diagram. The network of cross-references has been kept throughout this edition.

In compiling a dictionary of this kind, it becomes necessary to draw upon the work of many authorities and seek the advice of colleagues to all of whom the authors are deeply indebted.

Finally the authors express their sincere thanks to the publishers and printer for printing this book promptly.

All comments from users on omissions or shortcomings will be most welcome.

Dr. V.K. Puri
Dr. P. Chand

A

Ab : It is a prefix to a unit which indicates its application in the obsolete CGS electromagnetic system of units.

ABC : It is the abbrevation used for automatic brightness control.

Aberration : The defect in the image formed by an electronic lens system is called aberration.

Absolute Unit : When a quantity y, like charge or voltage, is completely represented by a simple function of the quantities such as x_1, x_2..(Where x may be time, current, etc.) in such a way that

$$y=f(x_1, x_2...)$$

and the unit of y, u_y, can be got from the units of x_1, x_2, etc., by the following equation

$$u_y \propto f(u_{x1}, u_{x2}...)$$

in case the constant of proportionality is unity, the unit is known as absolute unit. It means all SI units are absolute.

Absolute Zero : The temperature at which the energy of random motion of the particles in a system at thermal equilibrium is zero is known as absolute zero.

Absorption : It refers to attenuation of a radiowave because of dissipation of its energy, as by the production of heat.

Absorption Loss : It is the magnitude of the absorption of a radiowave, which is generally expressed in nepers or decibels.

Accelerated Life Test : It is a form of life test of a circuit or device designed in such a way that the duration of the test is appreciably less than the normal expected life of the device.

Accelerating Electrode : It is that electrode which accelerates electrons in the electron beam of an electron tube.

Acceleration Voltage : Any voltage which produces acceleration of a beam of charged particles.

Accelerator : It refers to a machine used to accelerate charged particles or ions in an electric field in order to give high-energy beams for the study of nuclear structure and reactions. Here the magnetic fields are used to focus and determine the direction of a beam.

Large energies are obtained using machines having beams of particles if subjected to a series of relatively small accelerating voltages. It should be noted that the particles travel either in a straight line, as in linear accelerators, or in cyclic paths.

A.C./D.C. Receiver : A radio receiver which can be operated by alternating current supply or a direct-current supply.

Acoustic Feedback : It refers to undesired feedback of the sound output of an audiofrequency loudspeaker to a preceding part of a sound-reproduction system.

Acoustic Wave : A wave which is transmitted through a solid, liquid, or gaseous material due to the mechanical vibrations of the particles forming the material. The normal mode of propagation is longitudinal which means the direction of motion of the particles is parallel to the direction of propagation of the wave, and the wave therefore consists of compressions and rafactions of the material. The term 'sound wave' is sometimes used for waves having frequency falling within the audible range of the human ear i.e., from about 20 hertz to 20 kilohertz.

Acoustic Wave Device : It is a device which is used in a signal-processing system in which acoustic waves are transmitted on a miniature substrate in order to perform a wide range of functions.

Bulk acoustic waves are regarded as acoustic waves propagated through the bulk substrate material. The substrate material consists of a piezoelectric semiconductor like cadmium sulphide. The acoustic waves are produced from electrical signals as a result of the piezoelectic effect.

Surface acoustic waves are those waves which are propagated along the surface of a substrate. Here the associated electric field extends for a short distance out of the surface and may interact with the conduction electrons of a separate semiconductor placed just above the surface.

Action Current : It is very small current wave associated with nerve impulses.

Active : It denotes any device, component, or circuit which introduces gain or has a directional function.

Active Current. : Or active component, energy component, power component, in-phase component of the current. It means the component of the current which is in phase with the voltage, alternating current and voltage being regarded as vector quantities.

Active Load : It is a load which is formed from an active device, specificaly an MOS transistor. The active device is not used for its inherent gain, but the resistance of the device is used to form the load MOS transistors are particularly useful as active load devices as no extra processing stops are required to form resistors, during the manufacture of MOS integrated circuits.

Active Voltage : Or active component, energy component, power component, in-phase component of the voltage.

Active Volt-Amperes : It is the product of the voltage and the active current or of the current and the active voltage which is equal to the power in watts.

Activity : It is regarded as the ratio of the peak value of the oscillations in a piezoelectric crystal to the peak value of the exciting voltage.

Actuator : It is a device used to calibrate electronic equipment, in a control system, to convert and electrical signal into the appropriate mechanical energy.

Adcock Direction Finder : A radio direction finder having a number of spaced vertical aerials.

Adder : It is regarded as a circuit in a digital computer which performs mathematical addition. Mostly it contains several identical sections each of which add the corresponding bits of the two numbers to be added together with a carry digit from the preceding section and give an output corresponding to the sum of the bits and a carry digit for the next section.

Aerial : *Syn Antenna*. The part of a radio system which radiates energy into space (*transmitting aerial*) or receives energy from space (*receiving aerial*). An aerial together with its feeders and all its supports is known as an aerial system.

Aerial Array. : It is regarded as an arrangement of radiating or receiving elements spaced and connected in such a way to produce directional effects. By this great directivity and consequently large aerial gain can be produced by suitable design.

Aerial Current : It is the root mean square value of the current determined at a specified point in an aerial, mostly either at the feed point or at the current maximum.

Aerial Efficiency : The ratio of the power radiated by an aerial, at a particular frequency, to the total power supplied to it.

Aerial Feed Point Impedance : It is the impedance of an aerial at the point at which it is fed.

Aerial Gain : (*i*) (in transmission) It is the ratio of the power which must be supplied to a reference aerial compared to the power supplied to the aerial under consideration in order that they give exactly similar field strengths at the same distance and in the same specified direction. (*ii*) (in reception). It is the ratio of the signal power produced at the receiver input by the given aerial to the given by a reference aerial under similar receiving conditions and transmitted power.

Aerial Radiation Resistance : It refers to the power radiated by an aerial divided by the mean square value of the current at a given specified reference point on the aerial, generally the feed point or a current antinode. Further this resistance takes into account the energy consumed by the aerial system due to radiation.

Aerial Resistance : It is the total power supplied to an aerial divided by the mean square value of the current at a given specified reference point on the aerial, generally the feed point or a current antinode. This resistance takes into account the energy consumed by the aerial system due to radiation and other losses.

Air-Break : (Of a switch, circuit-breaker, etc.,). Having contacts that separate in air.

Air Capacitor : It is the capacitor, which uses air as the main dielectric.

Alcomax Tradename : It is a material used for permanent magnets due to of its exceptionally high coercivity. It is made of an alloy of iron, nickel, aluminium, cobalt, and copper.

Aligned Grid Value : It refers to a type of thermionic value in which the power-handling capacity of the value is increased by arranging for only a very small fraction of the total space current to be intercepted by the screen grid. It is generally achieved by placing the grids so that the conductors forming the screen grid are in the shadow of those forming the control grid. Sometime special plates can be used to split the beam into thin pencils which are directed to fall mainly in the spaces between the grid conductors.

Alloyed Device : It is regarded as a semiconductor device, such as a transistor or diode which contains one or more alloyed junctions.

Alloyed Junction : It is regarded as a semiconductor junction obtained by bonding metal contacts on to a wafer of semiconductor material and then heating to produce an alloy. This method is commonly used for germanium diodes and transistors and in early silicon devices, the devices being termed *alloyed* (or *alloy*) *transistors* of *diodes*.

Alnico : Tradename. It is a material used for high-energy permanent magnets and is an alloy of nickel, iron, aluminium, cobalt, and copper.

Alpha Cut-off Frequency : It refers to that frequency at which the common base forward-current transfer radio, α of a bipolar junction transistor has fallen to 1√2 (*i.e.*, 0.707) of the low frequency value.

Alphanumeric : It is ordering or ordered by both letters and numbers.

Alpha Particle : (∝-particle) The nucleus of a helium (^{4}He) atom. The positively charged particle is very stable and consists of two protons and two neutrons.

Alpha Rays : A stream of alpha particles which have energies characteristic of the emitting radioactive substance.

Alternating-Current Motor : It is a motor that requires alternating current for its operation.

Alternating-Current Transmission : It refers to a method of transmission used in television in which the direct-current component of the luminance signal is not transmitted. Moreover a direct-current restorer must be used in this form of transmission.

Alumina : Aluminium oxide, symbol Al_2O_3. In case of solid-state electronics it is used as a dielectric, as in thin-film capacitors, or as the gate dielectric in MIS transistors. In case of valve electronics it is used as an insulator because of its excellent electrical and thermal resistance.

Aluminium : Symbol Al. It is a metal having atomic number 13 ad is used to a large extent in electric equipment and devices. It is ductile, malleable, resistant to corrosion, lightweight, easily evaporated on to surfaces. It is the element of choice in both micro and macroelectronic applications. It is a good conductor.

Aluminium Antimonide : Symbol : AlSb. It is a semiconductor having properties up to operating temperatures of 500°C.

AM/FM Receiver : It is a radio receiver which detects both amplitude-modulated and frequency-modulated signals.

Ammeter : An instrument which is used for measuring current. Some common types are moving coil and moving iron instruments and the thermoammeter. It should be noted that the most ammeters are shunted galvanometers.

Ampere : Symbol : A. The SI unit of electric current which is defined as the constant current which if maintained in two straight parallel conductors of infinite length and negligible cross section and placed one metre apart in a vacuum, would produce between these conductors a force having magnitude 2×10^{-7} newton per metre of length.

Ampere Per Metre : Symbol : A/m. It is the SI unit of magnetic field strength. It refers to the magnetic field strength in the interior of an elongated uniformly wound solenoid which is excited with a linear current density in its winding of one ampere per metre of Axial distance.

Ampere's Law : It states that the force between two parallel current carrying conductors in free space is given by

$$dF=\mu_0 I_1 ds_1 I_2 ds_2 \sin\theta/4\pi r^2$$

Syn. Ampere's circuital theorem. In case B is the magnetic flux density at a point in vacuo, due to current I, then the

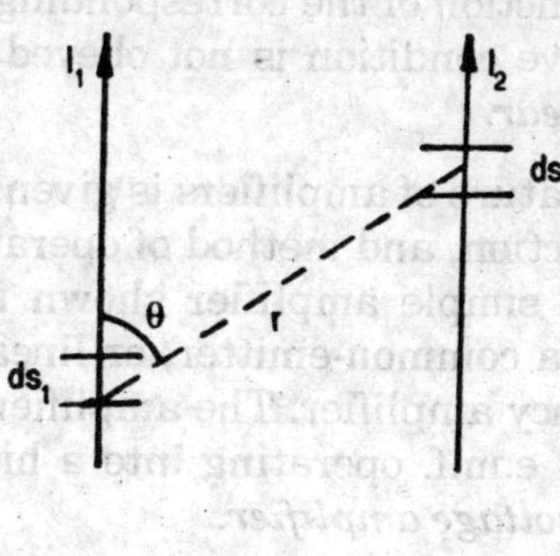

Ampere's law.

line integral of B over a closed circuit encircling/is represented by the equation

$$\phi B.ds = \mu_0 I$$

Where μ_0 is the permeability of free space and ds and incremental length.

The total current flowing is obtained by the integral of the current density flowing in the area bounded by the loop.

Ampere-Turn : symbol At. It is a unit of magnetomotive force which is equal to the product, NI, of the total number of turns, N, in a coil through which a current, I is flowing.

Amplification : The ratio of out put voltage to input voltage of current. This can be expressed mathematically as

$$A_\upsilon = \frac{C_{OUT}}{C_{in}} \text{ and } A_i = \frac{i_{OUT}}{i_{in}}$$

The symbol Av represents voltage amplification, while Ai stands for current amplification.

Amplification Factor : Symbol : μ. It is the ratio of the incremental change in output voltage required to maintain the constancy of the current to a corresponding incremental change in the input voltage. u is equal to —Av, where Av is the voltage gain of the device.

Amplifier : A device which gives an electrical output which is a function of corresponding electrical input parameter and increased the magnitude of the input with the help of energy drawn from an external source. A *linear amplifier* is one in which the instantaneous output signal is a linear function of the corresponding input signal. In case the above condition is not obeyed the amplifier is called *nonlinear*.

The nomenclature of amplifiers is given by their application, construction, and method of operation according to context. The simple amplifier shown in Fig. *a* can be regarded as a common-emitter, or linear, or class A, or audiofrequency amplifier. The amplifier which produces an increased e.m.f. operating into a high impedance is known as a *voltage amplifier*.

A transistor in common-base connection (Fig. *b*) acts as a

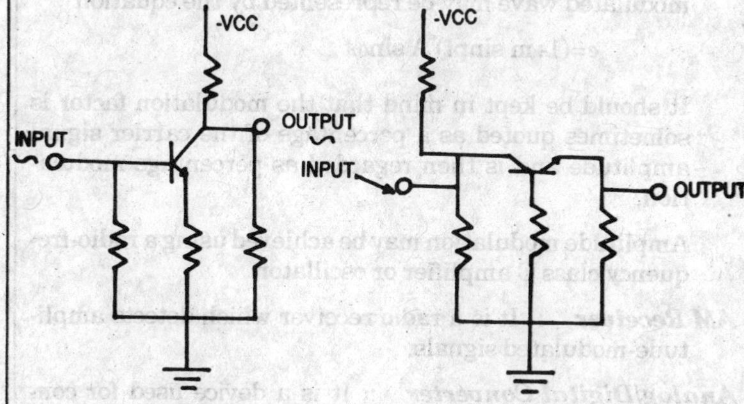

a **Common-emmitter n-p-n transistor** *b* **Common-base n-p-n trnsistor**

simple voltage amplifier operating with low input impedance and high out put impedance. An amplifier which gives an appreciable current flow into a relatively low impedance or a large increase in output power is a *power amplifier*.

Amplifier Stage: : It refers to a relatively small-gain amplifier which is coupled to other similar devices in cascade. In other words the output of one stage forms the input of the next, in order to provide a large overall gain. The *stage efficiency* is regarded as the ratio of the useful power output to the load of the amplifier stage, to the power supplied at the input.

Amplitude : (*i*) It is the peak value of an alternating current or wave in the positive or negative direction. The difference between extreme values in a complete cycle is the peak-to-peak amplitude.

(*ii*) It is value of an alternating current or wave in the positive or negative direction at a particular moment.

Amplitude Modulation (a.m. or AM) : It refers to a type of modulation in which the amplitude of the carrier wave is changed above and below its unmodulated value by an amount proportional to the amplitude of the signal wave and at the frequency of the modulating signal. In case modulation factor, m, is defined by m=B/A then the modulated wave may be represented by the equation

$$e=(1+m \sin pt)\ A \sin\omega t$$

It should be kept in mind that the modulation factor is sometimes quoted as a percentage of the carrier signal amplitude and is then regarded as percentage modulation.

Amplitude modulation may be achieved using a radio-frequency class C amplifier or oscillator.

AM Receiver : It is a radio receiver which detects amplitude-modulated signals.

Analog/Digital Converter : It is a device used for converting a continuously changing signal, like voltage or frequency, into a series of numbers suitable for use by digital equipment, such as a computer.

Anderson Bridge : It refers to a bridge, in which an inductance, L, is compared with a capacitance, C. In this case a null point detector instrument I, R_4 and X are adjusted until, when the bridge is balanced we have

$$R_2R_3=R_1R_4$$
$$L=C[R_2R_3+(R_3+R_4)X]$$

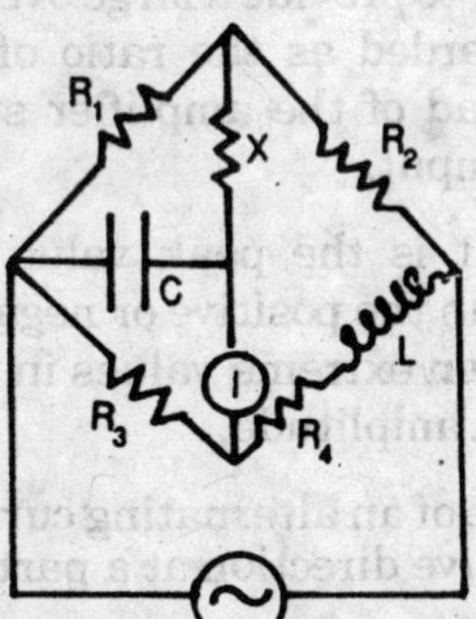

Anderson Bridge

Angle of Flow : It means the portion of the cycle of an alternating voltage, expressed as an angle, during which current flows.

Angstrom : Symbol : A. It is a unit of length equal to 10^{-10} metre.

Angular Frequancy : Symbol : W. It is defined as the frequency of a periodic phenomenon expressed in radians per second. It is equal to the frequency in hertz time 2.

Anion : It refers to an ion which carries a negative charge and, in electrolysis, moves towards the anode.

Anisotropic : It denotes a crystalline material whose properties, including conductivity, permittivity, and permeability, change with direction relative to the crystal axs as a result of the crystal structure.

Anode : It refers to the positive electrode of an electrolytic cell, discharge tube, value, or solid-sate rectifier i.e., the electrode by which electrons leave (and conventional current enters) a system.

Anode Current : It is the current flowing from the anode to the cathode of a device like solid-state rectifier.

Anode-Voltage Stabilized Camera Tube : Syn for high electron velocity camera tube.

Anticapacitance Switch : It is a switch which is designed in such a way so as to present the minimum possible series capacitance to a circuit when in the open position.

Anticathode : The term used for target electrode.

Anticoincidence Circuit : It is a circuit having two or more input terminals which produces an output signal when a signal is received by only one input. Moreover there is no output when a signal is received by each input either simultaneously or within a specified time interval, t. An electronic counter having this type of circuit will record single events and is called an anticoincidence counter.

Antiferromagnetism : This is the effect seen in certain solids whose magnetic properties change at a certain characteristic temperature, TN, called the Neel temperature.

Antihunting Circuit : It is a circuit designed in such a way so as to prevent oscillation in a feedback control loop thereby stabilizing it.

Anti-Interference Aerial System : It refers to a receiving aerial system designed in such a way that only the aerial itself abstracts energy from the incident radiation and no energy is collected by any feeders or supports of the system. The effects of interference can be minimised by locating the aerial remotely from sources of am-made interference, like electric motors.

Aperiodic Circuit : It is a circuit having both inductance and capacitance but is not capable of resonance as the total energy loss in the circuit is more than the critical value above which resonance does not occur. In other words the damping of the circuit is more than critical damping.

Apple Tube : This is a colour picture tube having the three colour phosphors arranged as narrow vertical stripes on the television screen.

Arbitrary Unit : A unit defined in terms of a prototype. The example is kilogram.

Arc : It is a luminous electric discharge in a gas having high current density and low potential gradeint.

Arc-Back : It is an arc struck from the anode to the cathode in a gaseous rectifier value (i.e., in the reverse direction). Thus arc-backs generally occur under fault conditions.

Arc Converter : It is a device used for generating alternating current from an electric arc.

Arcing Contacts : These are auxiliary contacts in any type of circuit-breaker switch which operate in conjunction with the main contacts, closing before and opening after them, in such a way as to protect them from arc damage.

Arcting Horn : It refers to a horn shaped conductor fitted to an insulator in order to prevent damage to the latter in the event of a power arc arising from fault conditions.

Arc Lamp : The lamp which uses the brilliant light accompanying an arc as a source of illumination. The colour of the light given out will depend on the gas inside the discharge tube.

Arc Welding : It is a kind of welding in which an arc discharge is developed between the metal parts to be joined together and a filler rod of the same material. The heat produced by the arc causes the metals to melt at the site to be joined, and they thus fuse together.

Argon. : Symbol Ar. Anoble gas having atomic number 18. It is extensively used as the gas in gas-filled tubes.

Armature : (1) The rotating part of an electric generator or motor.

(2) Any moving part in electrical equipment which closes a magnetic circuit or which has a voltage induced in it by a magnetic field. The simplest example is the moving contact in an electromagnetic relay.

Arrhenius Equation : It is an equation which expresses the time rate of degradation, $R(T)$, of some device parameter as a function of operating temperature, T, *i.e.*,

$$R(T) = A \exp(-eA_E/kT)$$

where e is electronic charge and k is the Boltzmann constant. The constant A_E is a measure of the average behaviour of a population of items and is the 'empirical' activation energy of the physical process causing component degradation. A is a constant and is the intercept of the plot of logarithm of degradation rate versus reciprocal temperature.

Arsenic : Symbol As. A metalloid having atomic number 33, that belongs to group VA of the periodic table. It is used as a donor impurity in semiconductors and in 3-5 compound semicouductors.

Artificial Aerial or Dummy Antenna : It means a device that simulates all the electrical characteristics of an actual aerial except that the energy supplied to it is not radiated. Its function is to allow adjustments to be made to a transmitter or receiver before connecting up to the actual aerial.

Artificial Line : It is regarded as an electrical network consisting of resistance, inductance, and capacitance which simulates the characteristics of a transmission line at any particular frequency.

Artwork. : It is the required patterns for integrated-circuit manufacture produced in a form suitable for reduction to masks. The pattern is due to alternate clear and opaque sections, generally in a form of Mylar film.

Aspect Ratio : it is the ratio of the width of a television picture to the height.

Assembly : A number of electronic components, parts, devices, or subassemblies (or a combination there of that are electrically and mechanically connected and that perform a specific function. When used, together with other assemblies, as part of a piece of electrical equipment the assembly may be removed or replaced as a whole, allowing easy maintenance and repair of the electrical equipment.

Astatic Galvanometer : It is regarded as a very sensitive type of moving-magnet Galvanometer which has two magnetic needles, NS and N′S′, made as nearly equal as possible and suspended on the same axis opposite direction.

Asymmetrical : It refers to a periodic quantity having a direct component.

Atom : An atom is the smallest particle of matter into which an element can be subdivided while retaining its original properties.

Atomic Number : Symbol : Z. The number of electrons moving around the nucleus of an atom, which is equal to the number of protons in the nucleus.

ATR Switch : It is the *abbrev.* used for anti-transmit-receive switch.

Attenuation : It refers to the reduction in magnitude of any electrical parameter of a signal, particularly electromagnetic radiation, on passing along any transmission path.

Attenuation Constant or Attenuation Coefficient : Symbol : x. It is the rate of exponential decrease in amplitude of voltage, current, or field component in the direction of propagation of a plane progressive wave, at a given frequency. When I_2 and I_1 are the currents at two points a distance d apart (I_1 being nearer the source of the wave), then we have

$$I_2 = I_1 \exp(-\propto d)$$

$\propto$ is mostly expressed in nepers or decibels.

Attenuation Distortion. : It is *syn*, with frequency distortion.

Attenuation Equalizer : It is that network which throughout a specified frequency band gives compensation for attenuation distortion.

Attenuator : A network or transducer designed to produce distortionless attenuation of an electrical signal. It may be variable or fixed. Attenuators are mostly calibrated in decibels.

Attracted-Disc Electrometer or Absolute Electrometer : It is an electrometer which measures potential difference in terms of fundamental mechanical quantities.

Audio Device : It is the abbrev. *for* audiofrequency device.

Audio Effect : It is the abbrev. *for* audiofrequency effect.

Audiofrequency (a f) : It is that frequency to which a normal human ear responds. In actual practice the audible range extends from about 20 to 20000 Hz. Intelligible speech can be obtained in a communication system when a frequency range from about 300 to 3400 Hz. is reproduced; any frequency in this range is called a *voice frequency*. Any electronic device, such as an amplifier, choke, or transformer, operating in this range is called as an *audiofrequency* device or *audio device*; similarly any effect such as distortion, involving audiofrequencies is called an *audiofrequency* effect or *audio effect*.

Audiometer : It is an instrument used to measure both hearing loss due to deafness and the masking produced by noise.

Automatic Control : It refers to any system or device which carries out operations automatically in response to the output of other electronic devices, like sensing devices, computers, or discriminator circuits.

Automatic Frequency Control (a.f.c) : A device to

maintain automatically the frequency of any source of alternating voltage within specified limits.

Automatic Gain Control (a.g.c.) or Automatic Volume Control: It is a device used for holding the output volume of a radio receiver substantially constant despite variations in the input signal.

Automatic Noise Limiter : It is a device in a radio receiver designed to limit the effect of impulse noise.

Automatic Tracking : The process of holding a radar beam locked on target while the target range is being evaluated.

Automatic Tuning Control : It is a kind of automatic frequency control in a radio receiver which adjusts the tuning to the correct setting for a given received signal if the manually operated adjustment has been set only approximately for that signal. Further it holds the tuning at the correct setting despite any small drift in the components of the receiver.

Autotransductor : It is a transductor having the main current and control current being carried in the same windings.

Autotransformer : It is a transformer which has a single winding, tapped at intervals, instead of two or more independent windings. It means that part winding is common to both primary and secondary circuits. The voltage V_2 across a tapped section is related to the total applied voltage V_1 by the following relation.

$$V_2/V_1 \propto n_2/n_1$$

where n_2 is the number of turns in the tapped section and

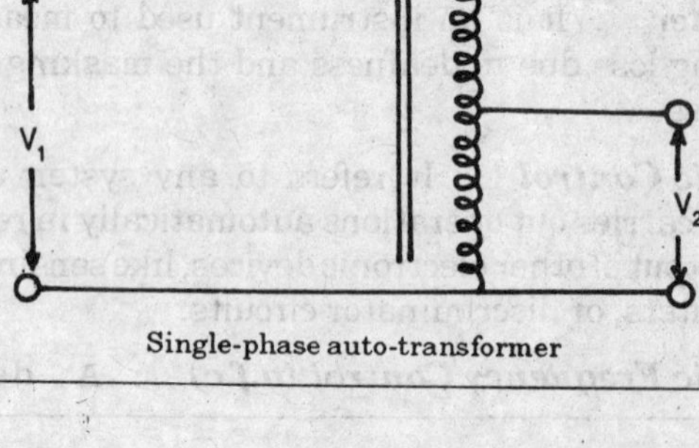

Single-phase auto-transformer

n_1 the total number in the winding.

Avalanche : It is the process of cumulative ionization in which a single particle or photon gives several ions, each of which in turn gains sufficient energy from an accelerating field to give more ions, and so on. It means a large number of charged particles is produced from the initial event.

Avalanche Breakdown : The breakdown which occurs in a semiconductor. It is due to the cumulative multiplication of free charge carriers under the action of an applied electric field.

The avalanche process takes place when a *critical field* is reached across the semiconductor. The critical field is the voltage gradient which just causes avalanche in a given specimen of semiconductor and is typical for the material used.

Avalanche breakdown is the breakdown which takes place in a reverse-biased p-n junction.

A.V.C : The Abbrev used for automatic volume control.

B

Back Electromotive Force : When e.m.f. opposes the normal flow of current in a circuit it is called back e.m.f. or back electromotive force normal flow of current in a circuit.

Background Counts : These are the counts registered by a radiation counter in the absence of the radiation source to be determined.

Backlash : It is regarded as the incomplete rectification of an alternating current in a thermionic value as a result of the presence of positive ions in the residual gas in the value.

Balance Controls : These may be defined in the following ways.

(*i*) Variable components used in electrical instruments and bridges in order to obtain conditions of electrical equilibrium in the circuit.

(*ii*) Variable components used to equalize the outputs of two or more similar circuits, like used in a stereophonic sound reproduction system.

Balanced Wire Circuit : It is that circuit whose two sides are symmetrical with respect to earth and other conductors and are electrically alike.

Ballast Resistor : It is a resistor manufactured from material having a large positive temperature coefficient of resistance.

Ballistic Galvanometer : It is that galvanometer which measures the quantity of electricity Q, flowing during the passage of a transient current I, where Q is defined by the following equation

$$Q = \int_0^\infty I dt.$$

The value of Q is determined from the deflection, θ of the moving part of the instrument. For a moving-coil instrument, $Q \propto \theta$ and for a suspended magnet type $Q \propto \sin 1/2$.

Balun : It is that device which is used to couple a balanced impedance, like an aerial, to an unbalanced transmission line, like coaxial cable. A balun is necessary in order to prevent asymmetrical loading of the balanced impedance and the induction of currents on the exterior of the unbalanced transmission line.

Banana Tube : It is that colour picture tube which has

three colour phosphors in the form of thin stripes on the screen. The line scan is formed inside the tube and the field scan is formed by cylindrical lenses mounted around the tube.

Band : It may be defined in the following ways.

(*i*) A specific range of frequencies used in communications for a definite purpose, such as the long-wave band in radio;

(*ii*) A closely spaced group of atomic energy levels.

(*iii*) A closely spaced group of molecular energy levels which appear as fluted bands separated by dark spaces in the spectra of compounds. When higher resolution is used, the bands resolve into fine spectral lines.

Bandwidth : It may be defined in the following ways. The band of frequencies occupied by a transmitted modulated signal (*see* modulation) and lying to each side of the carrier-wave frequency.

Bank : It may be defined in the following two ways

(*i*) A number of devices of the same kind, connected in such a way to act together.

(*ii*) An assembly of fixed contacts that is used in automatic switching in telephony to form a rigid unit in a selector or similar device with which wipers engage.

Barkhausen Effect : The effect which is observed in ferromagnetic materials where by the magnetization of the specimen proceeds as a series of finite jumps when the magnetizing flux is increased steadily.

Barrage Reception : It is a reception method used in telecommunications in which the receiving aerial consists of an array of several directive aerials of different orientations. The received signal is input from selected aerials in the array which are selected in such a way so as minimize interference from a particular direction.

Barretter : It is a device made of a sensitive metallic resistor (which is generally enclosed in a glass bulb) with a positive temperature coefficient of resistance.

Base : It is regarded as the region in a bipolar junction transistor between the emitter and collector into which minority carriers are injected.

Basic Frequency : In any oscillatory signal composed of several different sinusoidal components, the frequency of the component considered is known as basic frequency.

Batch Processing : It is process of operation employed with a large computer system in which a number of previously prepared programs are collected together and input to the computer as a unit.

Battery : A source of direct current or voltage which is made of two or more electrolytic cells connected together and used as a single unit.

A *dry battery* is a small portable battery made up from dry cells.

Baud : It refers to a unit of telegraph signalling speed equal to one unit element per second. For example in case the duration of the unit element is I/n seconds then the speed of transmission of successive signals in n bauds.

Bayonet Fitting : It is arrangement of pin-and-socket fitting in which the base of a lamp or tube has two pins, diametrically opposite each other, which can be inserted into the socket and rotated in slots in the socket in such a way that the device is held in position.

Beacon : It refers to a signal station or the signal transmitted by such a station, which acts as a reference point. A beacon which transmits an identifiable signal is known as a *code beacon*. A *homing beacon* is one that guides an objects, such as an aircraft, to a target, like an airport. At the airport, signals from the *localizer beacon* associated with the instrument landing system are picked up enabling the aircraft to be guided to land.

Beam : It refers to a narrow stream of unidirectional electromagnetic radiation or charged particles.

Beam Bending : An undesirable effect which takes place in television camera tubes. The electron beam used to scan the target area can be deflected from its intended position by the electrostatic charges stored on the target. This caused misalignment of the picture image in the receiver with respect to the original optical image.

Beam Coupling : It is the production in a circuit of an alternating current between two electrodes upon passage of an intensity-modulated electron beam. It should be noted that the *beam-coupling coefficient* is the ratio of the alternating current produced to the beam current.

Beam Lead : It is a connecting lead on a silicon chip which is made chemically and cantilevered across a void or apace on the chip.

Beat Reception : A method of radio reception which uses beating.

Beats : It is the periodic signal produced by interference in case two signals of slightly different frequencies are combined. Using a *beat-frequency oscillator*.

Beryllium Oxide : *Syn.* Beryllia. Symbol : BeO. It is regarded as an insulating material which has a high thermal conductivity and is used in heat sinks.

Beta Current Gain Factor Or Common-Emitter Forward-Current Transfer Ratio. Symboli : β or hfe.

Beta Decay : It is regarded as the spontaneous transformation of a nuclide into one of its isobars with the emission of an electron plus antineutrino or a positron plus neutrino. The mass of the nucleus is unchanged but the atomic number is changed by + 1 or – 1.

Beta Particle : (β-particle) An electron or positron emitted by a radioactive nucleus during beta decay.

Betatron : A cyclic accelerator which utilizes magnetic induction to produce high-energy electrons.

Bias : *Short for* bias voltage. A voltage applied to an electronic device to make sure that it operates on a particular portion of its characteristic curve.

Bifilar Suspension : It is an instrument in which the movable part is suspended by two threads, wires, or strips, arranged in such a manner that the restoring force is produced mainly by gravity.

Bifilar winding : In this method winding consisting of two contiguous insulated conductors are connected in

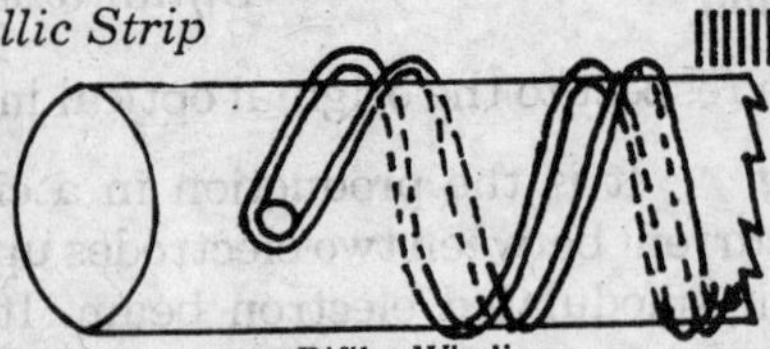

Bifilar Winding

such a manner they carry the same current in opposite directions. It gives a negligible magnetic field. The technique is mostly utilized for non-inductive resistors (*see diagram*).

Bimetallic Strip : In this device two metals with different coefficients of expansion are riveted together; an increase in temperature results strip to bend. One end of the strip is held rigidly and the movement of the other end may be used to open and close contacts, specifically in thermostats, or to move the pointer of a pointer-type thermometer.

Binary Notation : It refers to a process of numerical representation with two as the base and thus having only two digits (0 and 1). These symbols may be easily represented electronically by two voltage levels in a circuit and for this reason binary notation is used in computers.

Binding Energy : Symbol : E_B. 1. It can be defined in the following manners : The total energy released when protons and neutrons bind together to form a nucleus.

2. The energy required to release an extranuclear electron from a nucleus. It should be noted that the energy required to remove all the extranuclear electrons from a nucleus is the *total electron binding energy*.

Biot Symbol Bi : It is a unit of current in the obsolete CGS electromagnetic system of units and is equal to 10 amperes.

Biot-Savart Law : According to this law the magnetic flux density, B, at a point which is at a distance r from a current -carrying conductor in free space when a current I flows in an element of length ds is given by the relation

$$dB = (\mu_0 I ds \times r)/4\pi r^3$$

$$= \mu_0 I ds \sin\theta/4\pi r^2$$

Where μ_0 is the permeability of free space (*see* diagram).

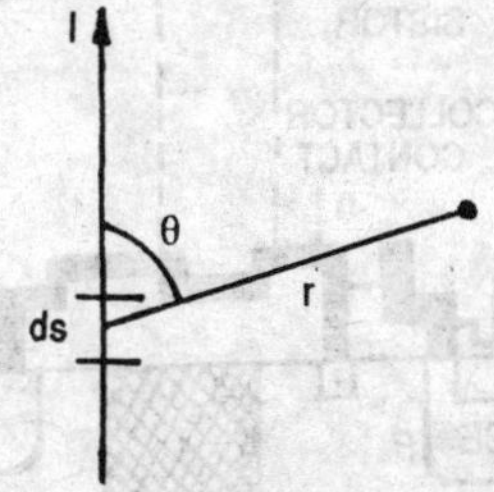

Biot-Savart law

It should be noted that the integration of the Biot-Savart law gives Ampere's circuital theorem.

Bipolar Integrated Circuit : It is a kind of monolithic integrated circuit based on bipolar transistors. A section of a typical circuit is shown in the diagram. The substrate (E) is obtained from a wafer of semiconductor (p-type is shown) and has buried n+ (highly doped) regions selectively diffused into it. These regions help in reducing the collector. Series resistance of the completed transistors. The n-type epitaxial layer (D) is then grown on the substrate to a specified thickness of 7 μm and this turn has an insulating oxide layer (B) grown on to it. Photolithography is employed to etch the oxide layer in the desired positions and isolating diffusions (C), of the same semiconductor type as the substrate, are made to isolate the individual components from each other. It should be noted that the individual components are formed by oxide growth, photoetching, and diffusion of the appropriate type of impurity, in turn, into the epitaxial layer. Finally a passivating oxide layer is grown. This has windows etched into it to help contacts to be made to the semiconductor; the desired interconnection pattern is made by evaporating a metal layer (A), generally aluminium, on to the oxide and etching.

Bipolar Transistor : It is a transistor like a junction transistor in which both electrons and holes play a necessary role in its operation. In common usage a bipolar junction transistor is generally known as a *transistor*.

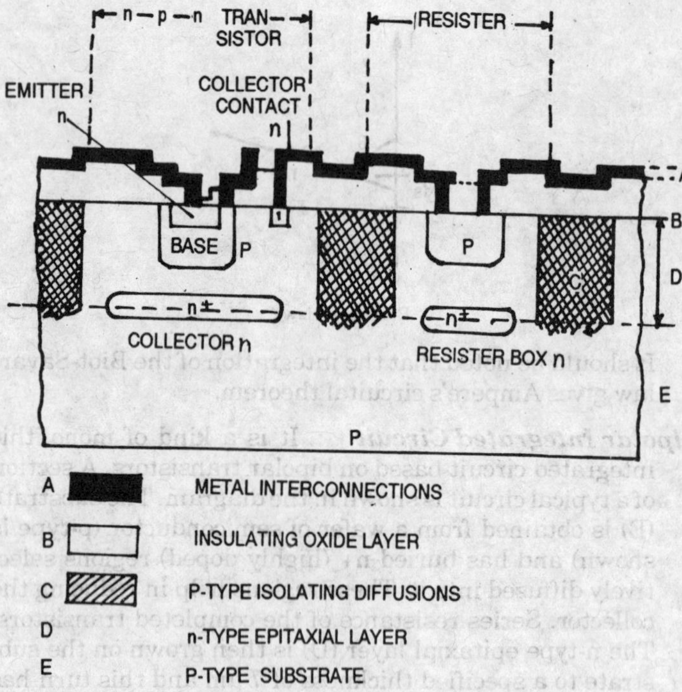

Cross section of typical bipolar integrated circuit

Bistable : *Short for* bistable multivibrator. It refers to circuit having two stable states.

Bit : Acronym from binary digit. One of the digits 0 or 1 used in binary notation. It is regarded as basic unit of information in a computer or data processing system.

Bitter Patterns : These are the experimental patterns demonstrating the presence of magnetic domains in ferromagnetic and ferrimagnetic materials.

Black Box : Any self contained unit or part of an electronic device which may be treated as a single package. This type of circuit may be approached mathematically in terms of its input and output characteristics, irrespective of its internal elements.

Black-Out : It is regarded as a temporary loss of sensitivity of any electronic device following the passage of an intense transient signal.

Blanking : Making a device or channel ineffective or inoperative for a desired time. Blanking eliminates the return trace from the screen of a cathode-ray tube. It should be noted that these are generally two blanking components to eliminate the horizontal and vertical components of the return trace.

Blasting : It is a form of distortion in audio and radio receivers as a result of extreme overloading of the amplifying circuits.

Block Diagram : It is a diagram in which actual circuits are represented by blocks as shown in Fig.

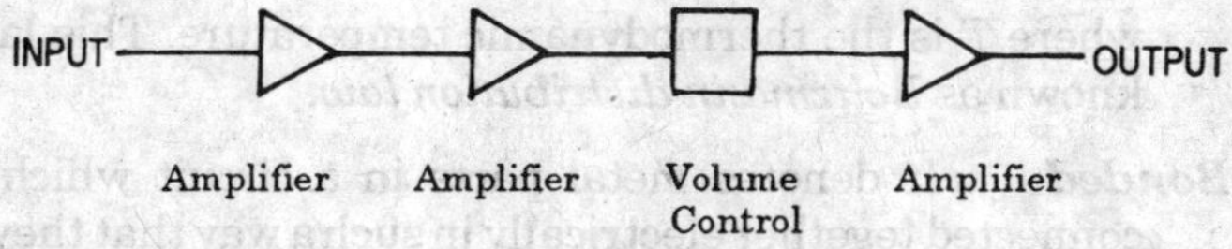

Blocked Impedance : It is the impedance measured when the mechanical motion of an electromechanical or acoustic transducer is blocked.

Blocking Capacitor : It is a capacitor included in a circuit in order to limit the flow of direct current and low frequency alternating current while allowing the passage of higher-frequency alternating current.

Blocking Layer Photocell : It is *Syn*. with barrier-layer photocell.

Blocking Oscillator : It is the oscillator in which blocking occurs after completion of (usually) one cycle of oscillation and lasts for a predetermined time. The whole process is then repeated.

Bode Diagram : A diagram in which gain or phase shift in a feed-back control system is plotted against frequency.

Bode Equalizer : An attenuation equalizer designed in such a manner that one simple control changes the amount of equalization in the same proportion for all frequencies within the range.

Body Capacitance : It is a capacitance obtained as a result of the proximity of a human body to a circuit.

Boella Effect : It is a reduction in the effective resistance or fixed composition resistors when operated at VHF of higher frequencies, due to dielectric losses.

Bolometer : It means a small resistive element capable of absorbing electromagnetic power. The resulting temperature rise is used to determine the power absorbed.

Boltzmann Constant : Symbol : k. It is a fundamental constant having the value $1.380\,41 \times 10^{-33}$ joule per kelvin.

In a system in which n_1 particles have an energy E_1 and n_2 particles have an energy E_2. We can write

$$n_1 = n_2 \exp[\,(E_1 - E_2)/kT]$$

where T is the thermodynamic temperature. This law is known as *Boltzmann distribution law*.

Bonded : It denotes metal parts in a circuit which are connected together electrically in such a way that they are at a common voltage.

Bonding Pads : Metal pads arranged on a semiconductor chip (generally around the edge) to which wires may be bonded in such a manner that electrical connection can be made to the component(s) or circuit(s) on the chip. Bonding is generally done by thermocompression or ultrasonic bonding.

Booster : It can be defined in two ways. 1. A generator or transformer inserted in a circuit in order to increase (*positive booster*) or decrease(*negative booster*) the magnitude or to change the phase of the voltage acting in the circuit.

A repeater station which amplifies and retransmits a broad-cast signal received from a main station, with or without a change of frequency.

Bootstrapping : It is as a technique used in a variety of applications in which a capacitor-the *bootstrap capacitor*-

is employed to provide 100% positive feedback for alternating currents across an amplifier stage of unity gain or less. Boot-strapping is used for control of the output signals by using the positive feedback to control the conditions in the input circuit in a desired manner.

Bootstrapping is generally employed in circuits that generate a linear time base, particularly in a sawtooth generator. A simple sawtooth generator consists of a capacitor which is charged by means of an input load resistor and discharged by a periodic step voltage. Since capacitor is charged, the voltage increases exponentially and as the voltage increases, that across the input load (and thus charging current) drops correspondingly. The output is approximately linear under the condition that only a small portion of the charging characteristic is used. The linearity may be further improved by using a bootstrap circuit to maintain a constant charging current. A typical circuit is given in Fig. (*a*). The output is mostly taken from an emitter follower, capacitively coupled via the bootstrap capacitor C_1 to the input load resistor R. As the output voltage rises, the voltage at the node between R and R_1 also rises; the voltage across R and thus charging current is therefore maintained substantially constant.

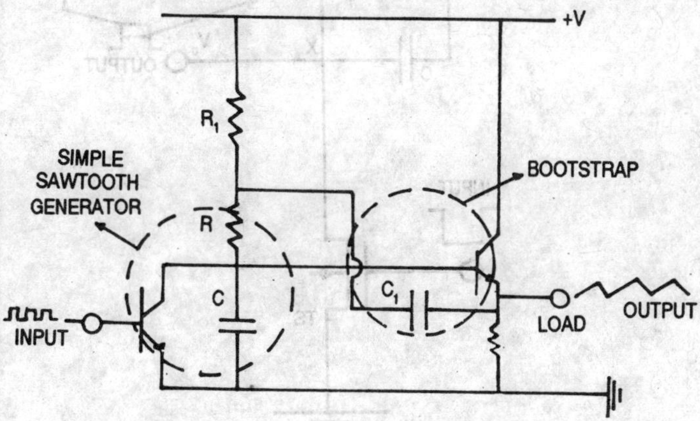

(a)—Bootsrapped sawtooth generator

If V_{G1} has a value greater than V_{DD}, then V_0 can also rise to a maximum possible value equal to V_{DD}. Bootstrapping is employed to achieve this effect. A bootstrap capacitor, *C*, is connected between points X and Y (Fig. *b*). As V_0 rises from the low logic level, V_{G1} also rises due to the positive feedback provided by *C*. It results in *TD* to switch off an point *B* is thus isolated from the supply bus. Thus V_{G1} can rise to a value greater than V_{DD} and as V_{A1} increases, V_0 also increases until it reaches V_{DD}. Hence the total voltage swing at *X* is optimized.

Bootstrapping may also be employed in high input impedance amplifiers, such as an emitter follower or a field effect transistor stage, in order to maximize the a.c. input impedance.

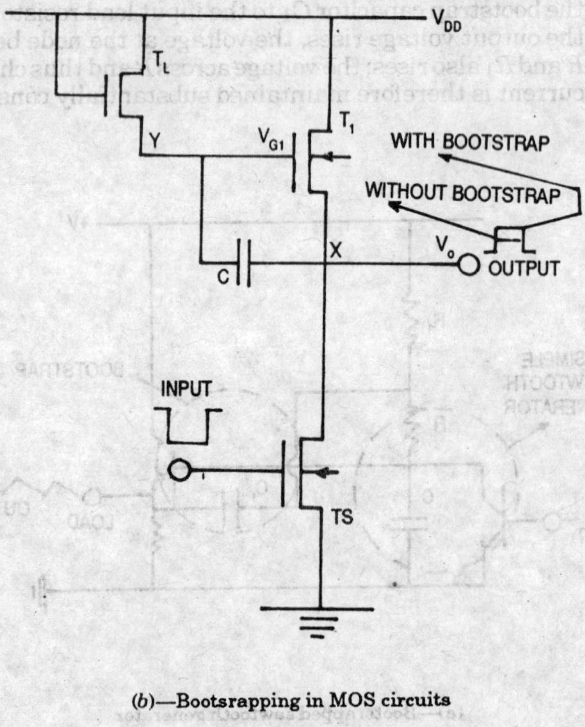

(*b*)—Bootsrapping in MOS circuits

Boron Resistor or Borocaron Resistor : A resistive film resistor which has a small percentage of boron introduced into the carbon film to make it more stable.

Bottoming : It refers to the operation of an electronic device in such a way that, when in the conducting state, the lowest instantaneous output voltage is determined by the device characteristics rather than the input voltage.

Breadboard Model : A rough assembly of discrete components which are generally attached temporarily to a board, for testing the feasibility of a circuit, system, or design principle.

Break : An accidental interruption of a broadcast programme.

Breakdown : It may be defined by only of the following definition.

1. A sudden catastrophic change in the properties of a device making it unfit for its purpose.

2. A sudden disruptive electrical discharge in an insulator or between the electrodes of an electron tube.

3. A sudden change from a high dynamic resistance to a much lower value in a semiconductor device.

Breakdown Voltage : (b.d.v.) It is the voltage under specified conditions at which breakdown occurs.

Bremsstrahlung : It is the electromagnetic radiation produced when an electron is suddenly decelerated by a nuclear field.

Bridge : It refers to one assembly of at least four circuit elements, like resistors, capacitors, etc., together with a current source and a null point detecting device (*see diagram*). Bridge networks from the basis of many measuring instruments.

Bridge Rectifier : It is a full-wave rectifier circuit in the form of a bridge, with 2 rectifier in each arm (*see diagram*).

Broadcasting : It means radio or television transmission to the public.

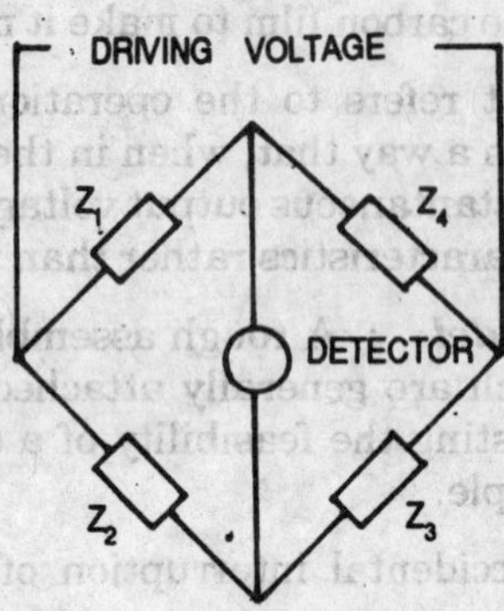

Bridge Circuit

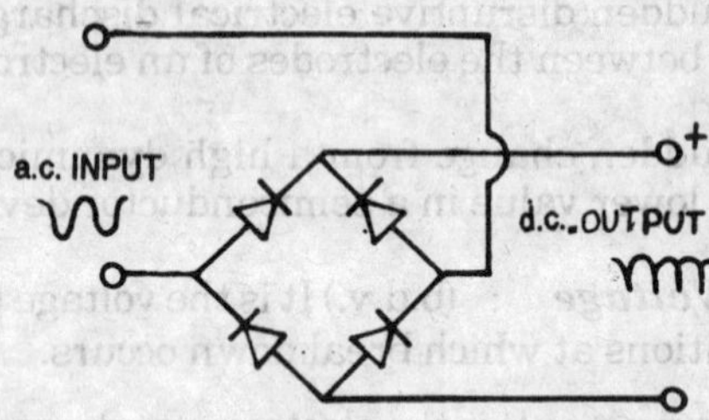

Bridge Rectifier Circuits

Brush : It is a conductor obtained from specially prepared carbon which provides electrical contact with a moving conductor, generally between the rotating and stationary parts of electrical machines. The contact resistance between the brush and the moving surface is the brush contact resistance.

Brush Discharge : It is regarded as a luminous electrical discharge which appears as a number of branching threads surrounding a conductor. Brush discharge takes place when the electric field around the conductor is not sufficiently large to produce a spark.

Bucket-Brigade Device (BBD) : A device which has a number of capacitors linked by a series of switches that in practice consist of bipolar or field-effect (MOS) transistors (see diagram). These circuits can be obtained from discrete components but are invariably manufactured as integrated circuits. Clock pulses are applied to close the switches, a two-phase system (ϕ_1 and ϕ_2) being used. As each switch is closed charge is transferred from one capacitor to the next.

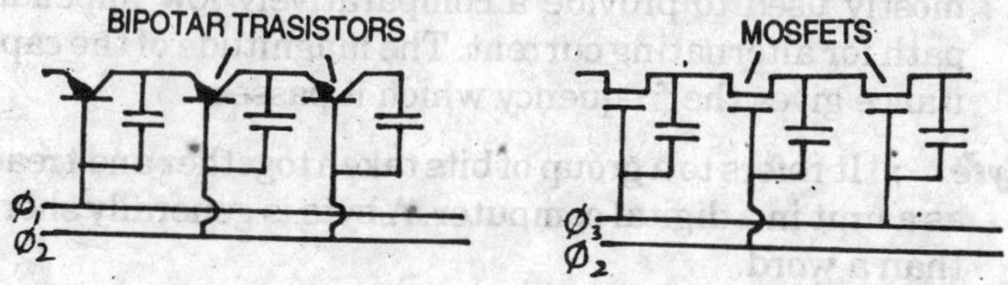

Bucket-brigade devices

It should be noted that the bucket brigade devices are mostly used as delay lines in both digital and analog systems because the amount of charge stored may change continuously from zero to the limit set by the magnitude of the capacitance and the operating voltage. The capacitors are generally used in practice by using the collector base capacitance or draingate capacitance of the transistors used in the circuit.

Buffer : It is an isolating circuit put between two circuits to minimize reaction from the output to the input. It may be used to handle a large fan out or to convert input and output voltage levels.

Bug : It refers to an error or fault in a computer program or in computer equipment.

Build-up Time : It is the time required for a current to rise to its maximum value in an electronic circuit or device.

Buried Layer : It refers to a layer of high conductivity semiconductor material diffused into the substrate layer while manufacturing bipolar integrated circuits and transistors.

Buried Signal Syn : It is *syn* with colour burst.

Bus or Busbar; Busline : It is a conductor having low impedance or high current-carrying capacity to which two or more circuits can be separately connected or which can connect several like points in a system, as with an earth bus. Busbars frequently feed power to various points.

Bushing : It is an insulator which is used to form a passage for a conductor through a partition.

By-Pass Capacitor : It is a shunt capacitor which is mostly used to provide a comparatively low impendace path for alternating current. The magnitude of the capacitance gives the frequency which it passes.

Byte : It refers to a group of bits taken together and treated as a unit in a digital computer. A byte is generally shorter than a word.

C

Cable : It is an assembly of conductors which has some degree of flexibility; the conductors are insulated from each other and enclosed in a common binding or sheath. The cables which are generally used are paired cable and coaxial cable although other configurations may be used. A uniform cable is one which has the same electrical characteristics at all points along its length.

Cable Track Locator : It is a device used to locate buried telephone cables. An audiofrequency oscillator, known as a bleeper, is used to send a signal along the cable in question and the bleeps are detected using a portable receiver and headset.

Calculator : It is a device that carries out logical and arithmetical operations of all types. The term is usually reserved for small machines capable of carrying out comparatively simple arithmetical tasks.

Calibration : It refers to the determination of the relation between the indicated value on a measuring instrument and the true value of the quantity to be measured. The true value is the value which would be obtained if all sources of error were eliminated.

Camera Tube : The device contained in a television camera which acts as an optical-electrical transducer and converts the optical image of the scene to be transmitted into electrical video signals.

Campbell Bridge : It means an a.c. bridge which is used to measure a mutual inductance, M, by comparing it with a standard capacitor, C. The resitances R_1 and R_2 are varied until null deflection is obtained on the indicating instrument I (see diagram). At balance point we have

$$L/M=(R+R_1)/R$$

$$M/C=RR_2$$

where L is the self-inductance of the coil AB.

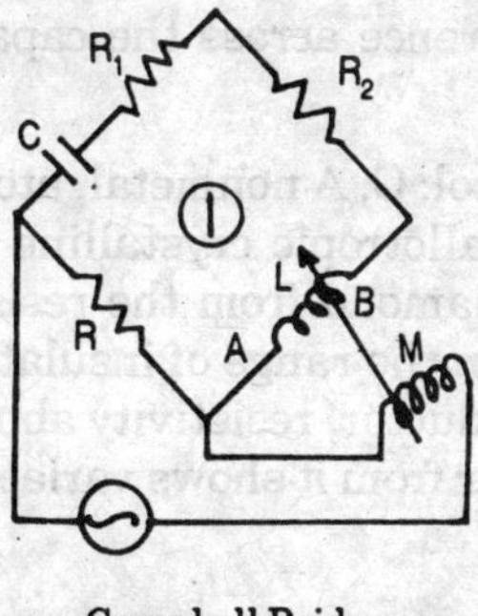

Campbell Bridge

Canned : It is a device which is completely enclosed and sealed by a metal sheath.

Capacitance : Symbol: C; Unit Farad. It is the property of an isolated conductor or a set of conductors and insulators whereby it stores electric charge. A charge of Q coulombs will increase the voltage of an isolated conductor by V volts. The capacitance is defined as the ratio Q/V and is determined by the size and shape of the conductor. It should be kept in mind that it is constant for a given isolated conductor.

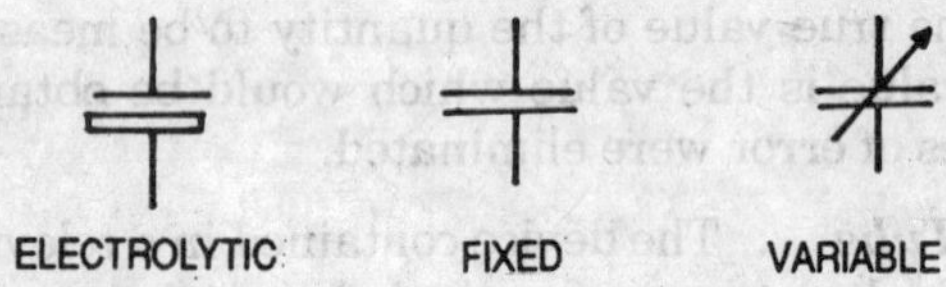

Capacitor Symbols

Capacitor : It refers to a component which has an appreciable capacitance. It consists of an arrangement of at least two conductors or semiconductors separated by a dielectric (an insulator). The conductors or semiconductors are termed as electrodes or plates. It should be noted that the value of the capacitance of a given device depends on the size and shape of the electrodes, the separation between them, and the relative permittivity of the dielectric.

Capacitor Microphone : It is a microphone in which a diaphragm forms one plate of a capacitor. Variations of sound pressure results in movement of the diaphragm and these alter the capacitance. Corresponding changes in potential difference across the capacitor are thus given out.

Carbon : Symbol: C. A nonmetal, atomic number 6, which exists in two allotropic crystalline forms: diamond and graphite. In diamond from the resistivity (5×10^{14} ohm cm) falls within the range of insulators. In graphite form it is a poor conductor, resistivity about 1.4×10^{-3} ohm cm, and in granular from it shows variation of resistance with pressure.

As graphite it is used to form resistors, in microphones, and as a filament of electric light.

Carbon Microphone: : It is a microphone which uses the variation of contact resistance of granular carbon with applied pressure.

Carcinotron : A crossed-field type of backward-wave oscillator.

Carey-Foster Bridge : It is a modified form of the Wheatstone bridge that measures the difference in resistance between two nearly equal resistances. The resistances are placed in the ratio arms of the bridge and the balance point obtained on a slide wire. The resistances are then switched and a new balance point obtained. The resistance difference is proportional to the distance between the balance points.

Carrier : A mobile electron or hole which transports charge through a metal or a semiconductor and is responsible for its conductivity.

Carrier Concentration : It refers to the number of charge carriers in a semiconductor per unit volume, in practice it is quoted as numbers per cubic centimetre.

In an extrinsic semiconductor the electrical neutrality of the sample is preserved and in a sample that contain impurities, we have

$$N_A^- + n = N_D^+ + p$$

where N_A^- and N_D^+ are the numbers of ionized acceptor and donor impurities, respectively. At relatively high temperatures most of the impurity atoms are ionized and therefore,

$$n + N_A \cong p + N_D$$

where N_A and N_D are the remaining numbers of non-ionized acceptors and donors. It should be noted that the product $np = ni^2$ is independent of added impurities.

In an *n*-type semiconductor the number of electrons, n_n, at thermal equilibrium is given by the reaction

$n_n \cong N_D$ if

$| N_D - N_A | > n_i$ and

$N_D N > N_A$

carrier concentrations | energy bands

The number of holes, p_n, in an n-type semiconductor is given by the relation

$$p_n = n_i^2/n_n$$

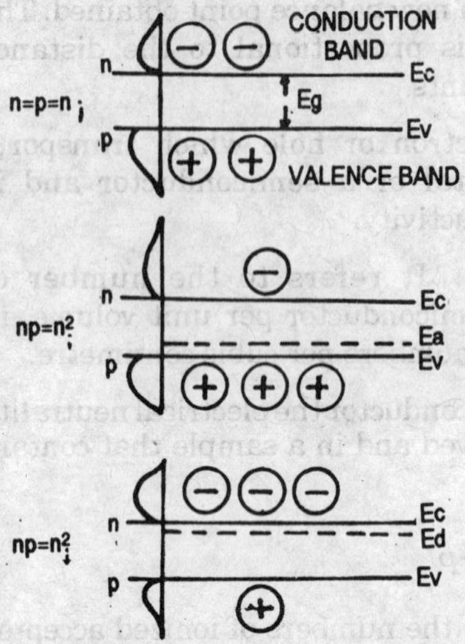

Intrinsic semiconductor

p-type semiconductor

n-type semiconductor

In a p-type semiconductor the situation is reversed :

$p_p \cong N_A$ if

$| N_A - N_D | >> n_i$ and

$N_A >> N_D$;

$$n_p = n_i^2/p_p$$

The carrier concentrations of i-, n-, and p type semiconductors are shown in the diagram.

Carrier Frequency : It refers to the frequency of a carrier wave.

Carrier Storage : An effect occurring in a *p–n* junction under forward bias.

Carrier System : It is transmission system used in telephony and telegraphy having modulated carrier waves transmitted along lines or cables. The multichannel carrier system in telephony permits many simultaneous independent signals to be transmitted on the same circuit.

Carrier Transmission : It is the transmission of a signal which is due to modulation of a carrier wave.

Carrier Wave : *Syn*. carrier. It is the wave that is used to be modulated in modulation, or, in a modulated wave, the carrier-frequency spectral component. The process of modulation produces spectral components falling into frequency bands at either the upper or lower side of the carrier frequency.

Cascade : It is the process of connecting in series a chain of electronic circuits or devices in such a way the output of one is the input of the next.

Cascade Control : If refers to an automatic control system in computer technology in which each control unit controls the succeeding unit and is controlled by the preceding one in the chain.

Catching Diode : It is a diode which is applied to limit the voltage at a point in a circuit. For this two diodes can be used to keep the voltage within specific limits (see diagram).

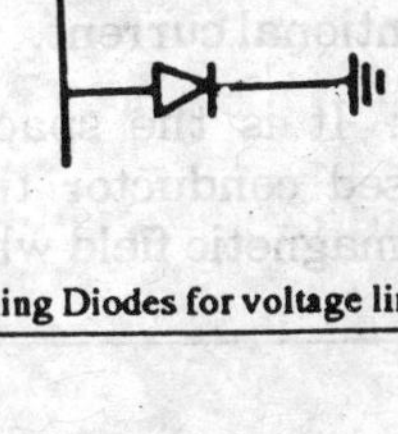

Catching Diodes for voltage limitation

Cathode : The negative electrode of an electrolytic cell, discharge tube, value, or solid-state rectifier is called cathode. It is also regarded as the electrode by which electrons enter (and conventional current leaves) a system.

Cathode Ray Oscilloscope (CRO) : It is an instrument in which a variety of electrical signals are presented on the screen of a cathode-ray tube for examination.

Cathode Ray Tube : It is a funnel shaped electron tube which converts electrical signals into a visible shape. All cathode ray tubes have an electron gun to generate an electron beam, a grid that varies the electron beam intensity and the brightness, and a luminescent screen to produce the display system. The electron beam is sent across the screen either by deflection plates or magnets.

Cathode-Ray Voltmeter : A voltmeter is a type of a cathode-ray tube of known deflection sensitivity.

Cathode Sputtering : It is *syn* of cathode disintegration. This is due to the slow disintegration of the cathode in a gas discharge tube due to bombardment by positive ions. The phenomenon is used for the deposition of thin metallic films on a surface. To achieve this the object to be coated is placed near the cathode of a gas-discharge tube, the cathode being made from the desired metal. The whole system is made as a demountable assembly that is evacuated to the correct pressure to form a gas tube.

Cathode-Voltage Stabilized Camera Tube : It is *syn* of low-electron velocity camera tube.

Cathodoluminescence : It is the emission of light while the substances are bombarded by cathode rays. The frequency of light emitted is property of the bombarded substance.

Cation : It is ion that carries a positive charge and, in electrolysis, moves towards the cathode, i.e., moves in the direction of conventional current.

Cavity Resonator : It is the space within a closed or substantially closed conductor that will maintain an oscillating electromagnetic field when excited externally.

The reasonant frequencies are determined by the size and shape of the cavity. The whole system has resonant effects and replaces tuned reasonant circuits for high-frequency applications for which the latter are impracticable.

CCD Filter : It is circuit in which the ability of a charged-coupled device to provide a precise predetermined delay time to an analong signal is utilized in order to produce a desired signal processing function.

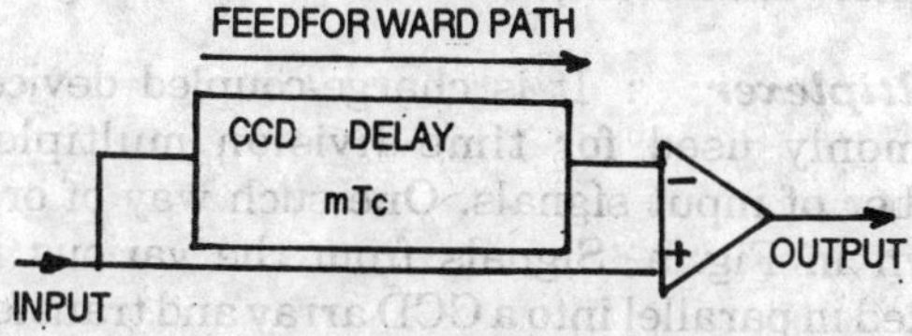

Fig. (*a*) Feedforward delay : simple transversal band-stop filter.

One type of the CCD filters is made by introducing a charge-coupled device of known delay into a circuit so that it provides a feedback (Fig. *a*) or feedforward (Fig. *b*) path. The circuits shown in Figs. *a* and *b* act as band pass and band stop (notch) filters, respectively. In both cases the signals that pass through the CCD array are subject to a constant deley time.

In multiple of half the input period, to give a band-pass filter. While in case of the circuit in Fig. *a* the signals

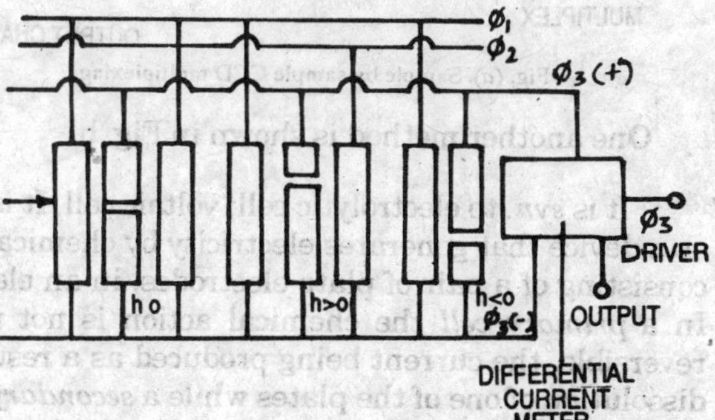

Fig. (*b*) schematic of 3-phase split-electrode transversal filter

are in phase if MTc is an integral multiple of the input period. There in phase signals generate a minimum output from the differential amplifier resulting in a bandscop filter.

The second main type of CCD filters uses the *split electrode technique* in order to produce a desired output. In this kind of transversal filter some of the gate electrodes of the CCD away are split into two parts is shown in (Fig. *b*). The charge packet is transferred through the array, the amount of charge which stored under the parts of the split electrode is in the ratio of the areas and the thickness is uniform, in the ratio of the lengths.

CCD Multiplexer : It is charge-coupled device which is commonly used for time-division multiplexing of a number of input signals. One such way of organization shown in Fig. a. Signals from the various inputs are clocked in parallel into a CCD array and transferred along the array. A similar array is used to separate the signals at the output. The Crosstalk between the signals can be reduced by using inert channels between each signal.

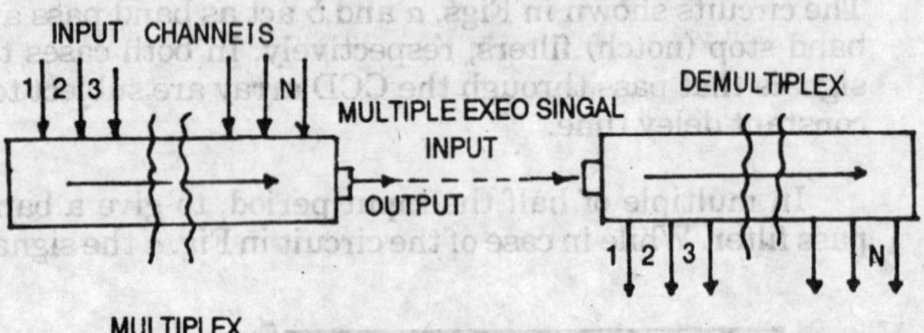

Fig. (*a*). Sample by sample CCD multiplexing.

One another method is shown in Fig. b.

Cell : It is *syn.* to electrolytic cell; voltaic cell. It is defined as a device that generates electricity by chemical means, consisting of a pair of plate electrodes. in an electrolyte. In a *primary cell* the chemical action is not normally reversible, the current being produced as a result of the dissolution of one of the plates while a *secondary cell* has reversible chemical action and is charged by passing a

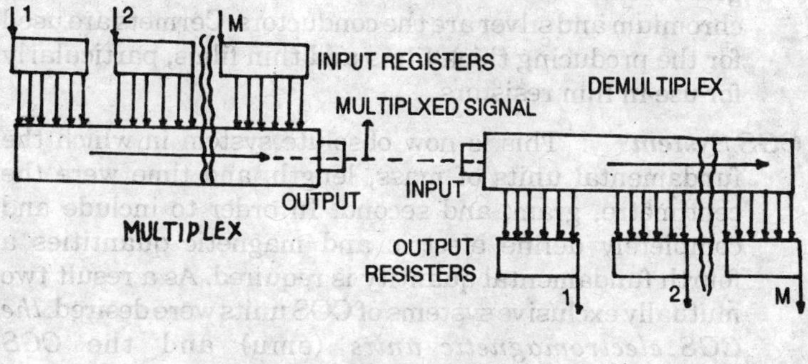

Fig. (*b*). Single segment CCD multiplexing.

current through it. The rate and direction of the chemical action is determined by the value of the external voltage.

A portable cell that has the electrolyte in the form of a nonspillable jelly or paste is called a *dry cell*. Most cells contain a liquid electrolyte and are sometimes termed wet cells. Dry cells are used in the batteries for torches, portable, radios, etc.

Cell Constant : It is defined as the area of the electrodes in an electrolytic cell divided by the distance between them. It is equal to the resistance in ohms of the cell when filled with a liquid of unit resistivity.

Centi : Symbol : c. A prefix to a unit, denoting a submultiple of 10^{-2} of that unit: one centimetre equals 10^{-2} metre.

Central Processing Unit (CPU) : It is the portion of a computer that controls the operation of the entire computer system and executes the arithmetical and logical functions contained in a particular program.

Ceramic Capacitor : It is defined as a capacitor utilizing a ceramic as the dielectric material. The behaviour of the capacitor is determined by the electrical properties of the ceramic used; these vary greatly but most have high permittivity allowing the capacitors to be smaller than most other types.

Cermet : It is a mixture of an insulating material and a highly conducting material. Ceramics like oxides and glasses are used as the insulator and metals like chromium and silver are the conductors. Cermets are used for the producing thick films and thin films, particularly for use in film resistors.

CGS System : This is now obsolute system in which the fundamental units of mass, length, and time were the centimetre, gram, and second. In order to include and completely define electric and magnetic quantities a fourth fundamental quantity is required. As a result two mutually exclusive systems of CGS units were desired: *the CGS electromagnetic units* (emu) and the *CGS electrostatic units* (esu).

Channel : A specified frequency band or a particular path used in communications for the reception or transmission of electrical signals.

Characteristic : A relationship between two values that characterizes the behaviour of a device, circuit, or apparatus is called characteristics. The relations area most commonly produced for unipolar and bipolar transistors, thermionic valves, and quadripoles. They are generally plotted in the form of families of graphs (*characteristic curves*) relating the currents obtained to the voltages applied for a range of operating conditions.

The *electrode characteristic* gives the relationship between current and voltage at an electrode of the device such as drain current versus drain voltage in a field-effect transistor.

Charge : Symbol : Q; unit: coulomb. Associated with each electron or proton is a tiny quantity of electricity which is referred to as charge. Electrons are assumed to carry a negative charge, while protons which carry opposite charge are said to be positively charged. A charged body has got an excess or lack of electrons with respect to its proton contents.

Charge Coupled Device (CCD) : It is a charge transfer device that consists of an array of MOS capacitors suitably designed so that they are coupled and therefore charges can be moved through the semiconductor substrate in a controlled manner.

Charge Density :

1. It is *syn.* to volume charge density. Symbol : ρ; unit: coulomb per metre cubed. It is defined as the electric charge per unit volume of a medium or body.
2. It is also *syn.* to surface charge density. Symbol: σ; unit: coulomb per square metre. It is defined as the electric charge per unit surface area of a medium or body.

Charge-Storage Diode : It is *syn.* to snap off diode; snapback diode; step-recovery diode. It is a *p-n* junction diode in which carrier storage is the major factor for the operation of the device. The diode is so designed that most of the injected minority carriers under forward bias are stored near the junction and are available for conduction as and when reverse bias is applied.

Charge-Transfer Device (CTD) : It is a semiconductor device in which discrete packets of charge are transferred from one location to the next. These types of devices are used for the short-term storage of charge in a particular location provided that the storage time is short compared with the recombination time in the material. The applications of charge-transfer devices include short-term memory systems, shift registers, and imaging systems.

Chip : This is a small piece of a crystal of semiconductor material containing either a single component of device or an integrated circuit. The chips are commonly sliced from a much larger wafer of substrate, in which many of components or circuits have been produced in chequerboard pattern.

Choke : It is *syn.* to choking coil. It is an inductor used to present a relatively high impedance to alternating current. The smoothing chokes are used to reduce the amount of fluctuation in the outputs of rectifying circuits. The smoothing chokes whose impedance varies with the current passing through them are called *swinging chokes*.

Chopper Amplifier : It is an amplifier that amplifies direct-current signals by first converting them into alternating current and then putting on normal a.c. amplifying techniques. This is achieved by using a system of relays or a suitable vibrator.

Chronotron : It is electronic device that gives the time interval between events. A pulse is initiated by each event and the time interval is determined by the position of the pulses along a transmission line.

Circuit : It is defined as the combination of a number of electrical devices and conductors that, when connected together to form a conducting path, fulfil a desired function like amplification. It is also defined as the path followed by negative charges (electrons) as they travel from negative end of a source of e.m.f. through a conductor and back to the positive side of source.

A circuit is *closed* when it forms a continuous path for current. When the path is broken, as by a switch, the circuit is open.

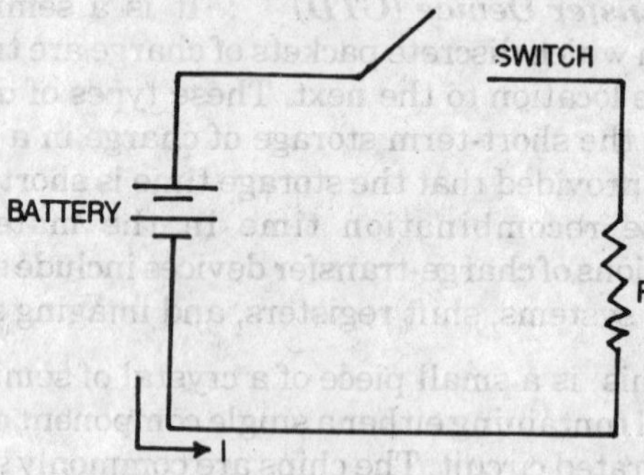

A circuit containing a source of e.m.f. a switch and a resistor

Circuit-Breaker : It is a device such as a contactor, switch or tripping device that is used to make or break a circuit under normal or fault conditions.

Circuit Diagram : It is a diagram that represents the function and interconnections of a circuit. Each circuit element is represented by its graphical symbol.

Circuit Parallel : It is one in which there are several paths for current to be travelling from one end of source of potential to the other.

Circuit Series : It exists when all the current from the source of e.m.f. has only a single path that it can flow in travelling from negative terminal of source through the circuit and to the positive terminal.

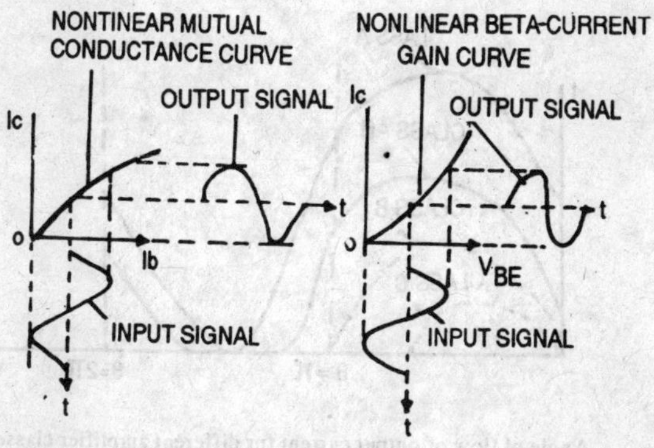

Distortion in class A amplifier with large-single operation

Class A Amplifier : It is a linear amplifier having output current flowing over the whole of the input current cycle, i.e. the angle of flow equal 2π. Such amplifiers have low distortion but low efficiency. The distortion can occur with large-signal operation due to the device transfer characteristics becoming nonlinear (*see* diagram above).

Class AB Amplifier : It is a linear amplifier having the output current flowing for more than half but less than the whole of the input cycle, *i.e.*, the angle of flow is between π and 2π.

Class B Amplifier : It is a linear amplifier operated so that the output current is cut off at zero input signal, i.e., the angle of flow equals π and a half-wave rectified output is produced.

Class C Amplifier : It is a nonlinear amplifier in which the output current flows for less than half the input cycle, *i.e.*, the angle of flow is less than π. Although more efficient than other types of amplifier yet class C amplifiers introduce more distortion.

Class D Amplifier : This is an amplifier operated by means of pulse-with modulation.

In theory class D amplifiers are very highly efficient but they require an impractically high speed of switching to avoid distortion.

Clear : It is a term used to restore a memory or storage device to a standard state, generally, the zero state.

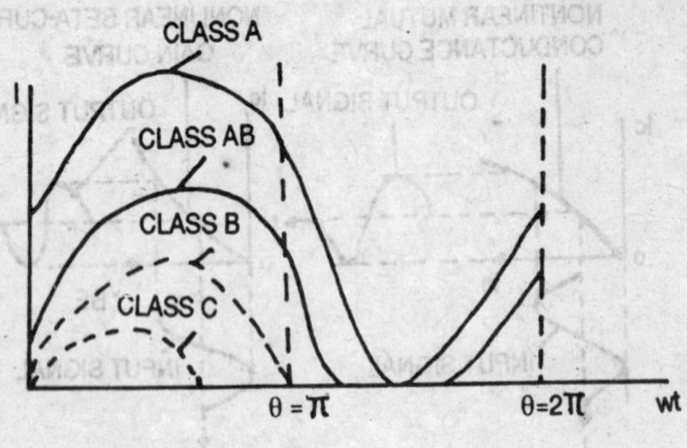

Angle of flow of output current for different amplifier classes

Clipper : This is *syn.* to voltage selector; peak limiter.

Clock : It is defined as an electronic device that generates periodic signals used to synchronize operations in a computer or to monitor and measure properties of the circuits involved.

Closed-Circuit Television : It is defined as a television system, other than broadcast television, that forms a closed circuit between television camera and receiver.

Clutter : Undesired signals that are caused by noise, unwanted images, or echos and appear on the display of a radar system.

CML : It is *abbrev.* for current mode logic.

Coated Cathode : It is *syn.* to Wehnelt cathode.

Coax : It is *short for* coaxial cable.

Coaxial Cable : It is defined as a cable formed from two or more coaxial cylindrical conductors insulated from each other. The outermost conductor is often earthed. Coaxial cables are extensively used to transmit high-frequency signals, as in television or radio, since they produce no external fields and are not affected by them. A terminal to which a coaxial cable may be connected is called a *coaxial terminal*.

Cockcroft-Walton Generator (or Accelerator). : It is a high-voltage direct current accelerator consisting of cascaded rectifier circuits and capacitors, to which a low-voltage alternating current is applied.

Coherent Detection : It is *syn.* for synchronous detection.

Coherent Radiation : Is radiation in which the waves are in phase both spatially and temporally.

Coil : It is a conductor or conductors wound in a series of turns. Coils are generally used to form inductors or the windings of transformers and motors.

Coincidence Circuit : It is a circuit with two or more input terminals which generates an output signal only when an input signal is received by each input either simultaneously or with a specified time interval, Δt. An electronic counter incorporating this type of a circuit records events occurring together is termed a coincidence counter.

Cold Cathode : A cathode from which field emission of electrons occurs at ambient temperatures due to the application of sufficiently high voltage gradient at its surface.

Collector : Short for collector region. It is defined as the region in a bipolar junction transistor into which carriers flow from the base through the collector junction.

Collector Current Multiplication Factor : It is the ratio of enhanced current flow in the collector of a transistor as a result of minority carriers entering from the base with sufficient energy to create extra electron hole pairs, to the current carried by minority carriers at the collector voltage. This ratio is generally unity but under high field conditions increases rapidly as the avalanche breakdown voltage is reached.

Colour Code : It is a method of marking electronic parts, such as resistors, giving information for the user. The value, tolerance, voltage rating, and any special characteristic of the component can be indicated using coloured bands or dots painted on it.

Colour Fringing : It is the presence of unwanted coloured fringes round the edges of objects viewed on the screen of a colour picture tube. Fringing is particularly noticeable when a monochrome signal is received.

Colour Picture Tube : It is defined as a cathode-ray tube

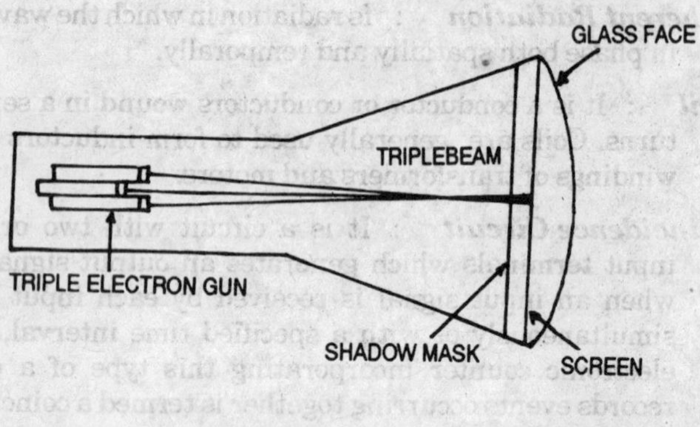

Fig.. *a.* Colourtron

designed to produce the coloured image in colour television. The coloured image is produced by varying the intensity of excitation of three different phosphors that produce the three primary colours red, green, and blue and reproduce the original colours of the image by an additive colour process.

The three gun colour picture tube has a configuration of three electron guns-the *red gun, blue gun,* and *green gun*-in other words then are titled slightly in such a way that the electron beams intersect just in front of the screen.

One main type is the *dot matrix tube*, for example colourtron has triangular arrangement of electron guns and has the phosphorus arranged as triangular sets of coloured dots (fig. a). A metal *shadow mask* is put directly behind the screen, in the plane of intersection of the electron beams, to ensure that each beam hits the correct phosphor (Fig. *b*).

The other main type of three-gun colour picture tube is the *slot matrix tube*, in this type the electron guns are arranged as vertical stripes on the screen (fig. *C*) and the shadow mask is replaced by an aperture grille of vertical wires.

The *Trinitron* colour picture tube has same advantages over three-gun tubes. It has a single electron gun having three cathodes aligned horizontally, an aperture grille, and vertically striped phosphorus. The cathodes are titled towards the centre in such a way that the electron beams intersect twice, one within the electron lens focusing system and once at the *aperture grille* (Fig. *d*).

The three-cathode arrangement of the trinitron allows a greater lens aperture than in the case of three-gun arrangement. The diameter of the electron tube for a given screen size is also reduced in the trinitron.

Cathode-ray tubes of the trinitron type can be used for applications where multiple electron beams are required; the angle of the tube can be increased to give a relatively wide-angle colour picture tube and thus a relative reduction in the overall size of colour television receivers.

The size of the colour cell is the smallest area on the screen which includes a complete set of the three primary colours. A smaller colour cell is available with horizontally aligned electron guns or cathodes than is possible in the triangular configuration.

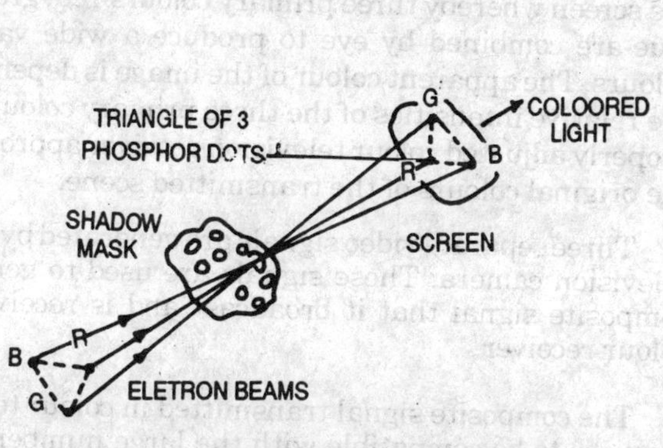

Fig. *b*. Light production in colourtron

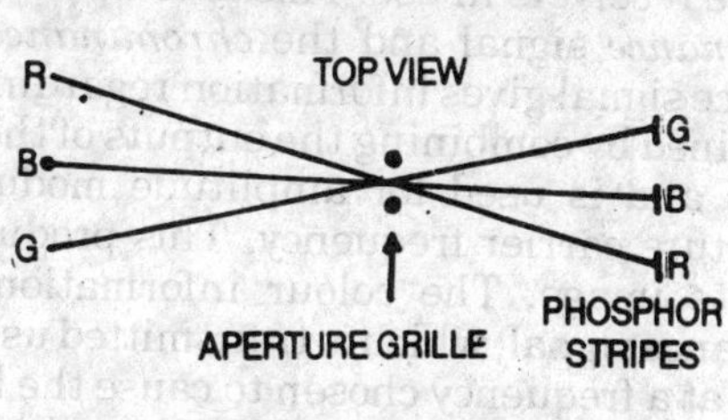

Fig. *c* Horizontal arrangement of electron guns

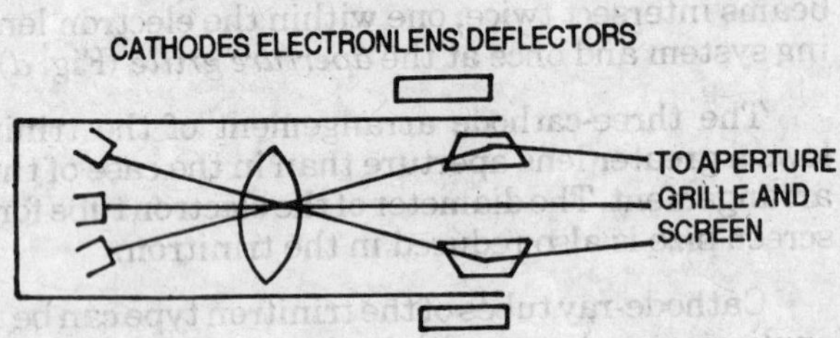

Fig. d. Trinitron

Colour Separation Overlay (CSO) : This is a technique used in colour television for superimposing part of one scene on another.

Colour Television : It is a television system that produces a coloured image on the screen of a colour picture tube. An additive colour reproduction process is used on the screen whereby three primary colours-red, green, and blue-are combined by eye to produce a wide variety of colours. The apparent colour of the image is dependent on the relative intensities of the three primary colours and a properly adjusted colour television receiver approximates the original colours of the transmitted scene.

Three separate video signals are generated by a colour television camera. These signals are used to generate a composite signal that it broadcast and is received by a colour receiver.

The composite signal transmitted in colour television requires to be compatible with the large number of back and white receivers in use. This is composed of two parts; the *luminance* signal and the *chrominance signal.* The luminance signal gives information regarding brightness. It is obtained by combining the outputs of the three colour channels and is used for amplitude modulation of the main picture carrier frequency. This produces the black and white image. The colour information is given by chrominance signal, which is transmitted using a subcarrier wave at a frequency chosen to cause the least interference on a monochrome set. The chrominance signal is received by combining, in a colour coder circuit, fixed specified fractions of the separate video signals into sum and difference signals. Two quadrature components of the chrominance signal are given and used for amplitude modulation of the *chrominance subcarriers*. The sub-carriers are suppressed at transmission. The original information is obtained from the chrominance signal in the *colour decoder* in the receiver. The frequency overlap is the range of transmitted frequencies which are same in both the luminance and chrominance channels.

Common-Base Connection : It is *syn.* to grounded-base connection. It is defined as a method of operating a transistor in which the base is common to both the input and output circuits and is generally earthed (see diagram). The emitter is used as the input terminal and the collector as the output terminal.

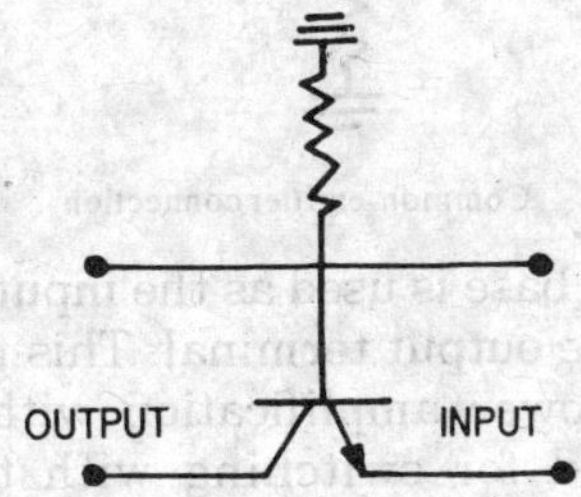

Common-Base Connection

Common-Base Forward-Current Transfer Radio : It is *syn.* to collector efficiency; alpha current factor; common base current gain symbol : α or h*fo*. It is defined as the ratio of the collector current I to the emitter current I_e of a bipolar junction transistor operated in common-base connection.

Common-Collector Connection : It is *syn.* to ground collector connection. This is a method of operating a transistor in which the collector is common to both the input and output circuits and is usually earthed (see diagram). The base is used as the input terminal and the emitter as the output terminal.

Common-Emitter Connection : It is *syn.* to ground emitter connection. This is a method of operating a transistor in which the emitter is common to both the input and output circuits and is generally earthed (see

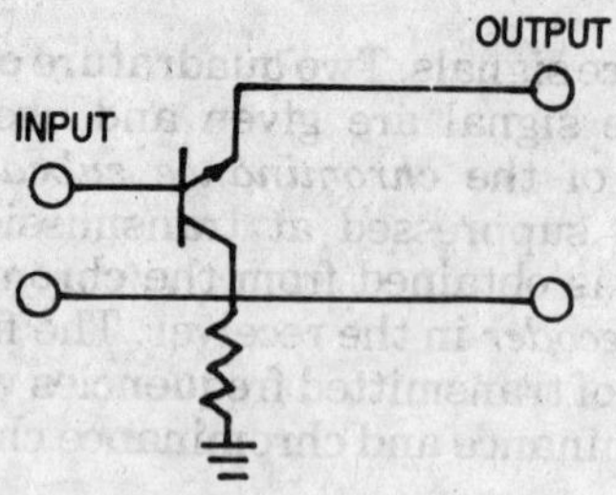

Common-collector connection

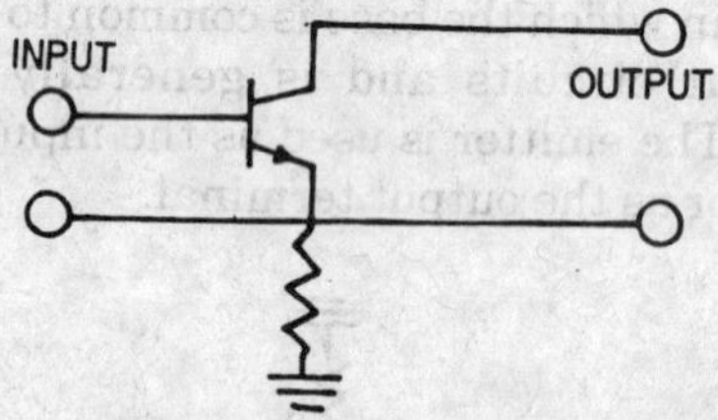

Common-emitter connection

diagram). The base is used as the input terminal and the collector as the output terminal. This type of connection is used for power amplification with a nonsaturated transistor and for switching with the transistor in saturation.

Communications Satellite (comsat) : It is defined as an artificial unmanned satellite in earth orbit which provides high-capacity communication links between widely separated locations on earth. International telephone services and the exchange or live television programmes and news are achieved by transmitting microwave signals, suitably modulated, from an earth station to an orbiting satellite and return to another earth location.

The first satellites, including the large metallized balloons *Echo* I and *Echo* 2, were passive systems; they simply reflected or scattered the microwave beam back to another earth station.

The satellite in a *geosynchronous* orbit revolves from west to east in an orbital period equal to that of the earth's rotation. it traces the path of figure-of-eight pattern in the sky with a latitude range depending on the inclination of orbit to the equatorial plane. For global coverage a minimum three geostationary satellites are needed, situated over the equatorial Atlantic, Indian, and Pacific Oceans.

Solar cells are the primary power supply for satellites with a back-up of bacteries for use during the brief periods of solar eclipses.

A geostationary satellite can be placed in a stable attitude by spinning it about an axis parallel to the earth's axis. In Intelsat IV the high-gain aerials are mounted on a plateform which rotates about the spin axis but in the opposite direction. The aerials then appear stationary with respect to the earth, at their desired orientation.

Commutation : It is defined as the transfer of current from one path in a circuit to a different one. The transfer is often done in a periodic and automatic manner.

Commutation Switch : It is defined as a switch that automatically switches from open to closed and vice versa repetitively. It is generally used for pulse modulation.

Comparator : It is a circuit, such as a differential amplifier, that compares two inputs and produces and output which is a function of the result of the comparison.

Compensating Leads : These are extra pair of leads that are used in conjunction with a sensitive measuring instrument connected to a bridge circuit.

Complementary Transistors : They are a pair of transistors of opposite type used together.

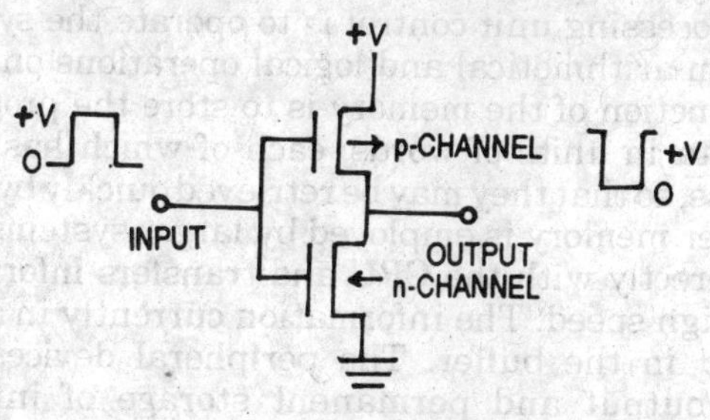

C/MOS inverter circuit

Complementary MOS field-effect transistors (*abbrev.* COSMOS or C/MOS) are generally utilized for low-dissipation logic circuits. This type of devices are used in conjunction with other similar stages all having essentially capacitive input impedance and therefore zero dc current flow. In the basic inverter (see diagram) the *p*-channel device conducts when the input is low and the *n*-channel device conducts when the input is high.

Component : It is defined as a discrete packaged electronic element, such as a resistor, that performs one electrical function.

Composite Conductor : It is a conductor formed from strands of different metals, such as copper and steel.

Composition Resistor : Generally used for a carbon-composition resistor.

Computer : It is the term used for any automatic device for the processing of information received in a prescribed and acceptable from according to a set of instructions. The instructions and informations are stored in a memory. The most widely used and most versatile of these devices is the digital computer, which can manipulate large amounts of information at high speed. Its input must be discrete rather than continuous and may consist of combinations of numbers, characters, and special symbols, the instruction-the program-being written in an appropriate programming language.

Most of the *computer systems* consist of three basic elements: the central processing unit (CPU), the main memory, and peripheral devices. The function of the central processing unit control is to operate the system and perform arithmetical and logical operations on the data. The function of the memory is to store the program and the data in units of words, each of which has a unique address, so that they may be retrieved quickly by the CPU. A buffer memory is employed by larger systems: it interacts directly with the CPU and transfers information at very high speed. The information currently in active use is held in the buffer. The peripheral devices perform 'input/output and permanent storage of information which is then converted into a set of discrete values for digital processing. The digital processing is considerably faster and more sophisticated than analog processing.

Concentration Cell : It is a type of cell having the electrodes made from the same metal and are immersed in two different concentrations of one salt of the metal.

Conductance : It is real part of the admittance, Y, which is given by equation

$$Y = G + iB$$

where B is the susceptance. For a direct-current circuit be conductance is the reciprocal of the resistance.

Conduction : It is transmission of electric energy through a substance that does not itself move. In electrical conductors, like metals, it entails the migration of electrons: in gases and solids it results from the migration of ions.

Conduction Band : It is defined as the band of energies of electrons in a solid in which the electrons can move freely under the influence of an electric field. In metals the conduction band is the highest occupied band.

Conduction Electrons : Electrons in the conduction beard of a solid.

Conductivity : Symbol : κ or σ and unit: siemens per metre. It is defined as the reciprocal of resistivity. It is given by the ratio of current density, j, to electric field strengtη, E.

$$k = j/E$$

This is an expression of Ohms law. A conductivity meter is an instrument used for measuring conductivity.

Conductor : It is a material that offers a low resistance to the passage of electric current: when a potential difference is applied across it a relatively large current flows.

Confusion Reflector : It is a device used to produce false signals with radar. Strips of paper or metal foil may be used for the purpose..

Conjugate Impedances : Two impedances given by equations,

$$Z_1 = R + iX$$

$$Z_2 = R - iX$$

in which the resistance components, R, are equal and the reactance components, X, are equal in magnitude but opposite in sign.

Constantan. : It is an alloy of copper and nickel that has a very low temperature coefficient of resistance and a comparatively high resistance.

Constant-Current Source : It is a circuit that has an infinietly high output impedance so that the output current is independent of voltage, practically sufficient high output impedance are only achieved for a limited range of output voltages.

Constant-Voltage Source : It is a source of voltage that produces a substantially constant value of voltage independently of the current supplied by it.

Contact : It is bringing together of two conductors so that current can flow. The resistance at the surface of contact is called the *contact resistance*. In case the conductors are made from two different materials a difference of potential will arise when they are placed in contact.

Control Electrode : It is defined as an electrode to which a signal is applied to produce changes in the currents of one or more of the other.

Convergence : It is defined as the intersection of the beams at a specified point. In a multibeam electron tube this may be obtaines electrostatically using a convergence electrode or electromagnetically using a convergence magnet.

Converter. : It is defined as a device for converting alternating current to direct current or vice versa.

It is also a device that changes the frequency of a signal;

This is also a device, such as an impedance converter, that has different electrical properties at its input and output and may be used to couple dissimilar circuits.

It is a device, such as a compiler, that changes an information code.

It is a transducer that converts energy of one type, into electrical energy such as sound wave to electro-magnetic radiation.

Coplanar Process : It is a technique used during the manufacture of LSI MOS integrated circuits and some forms of LSI bipolar integrated circuits, like I^2L. Regions of relatively thick silicon dioxide are used in order to isolate device areas and to prevent spurious MOST formation. The purpose of development of coplanar process is to minimize the vertical projection of the oxide layer. Silicon nitride layer is deposited on the surface of the silicon wafer and is etched to expose the regions of the surface where thick oxide is required. When oxidation takes place, the effective silicon surface moves downwards and is replaced by a thicker layer of silicon dioxide, so that approximately one third of the oxide is below the original exposed surface level (Fig. a). The oxidation may be preceded by etching of he exposed silicon surface in such a way the final oxide surface is at the same level as the original substrate (Fig.b).

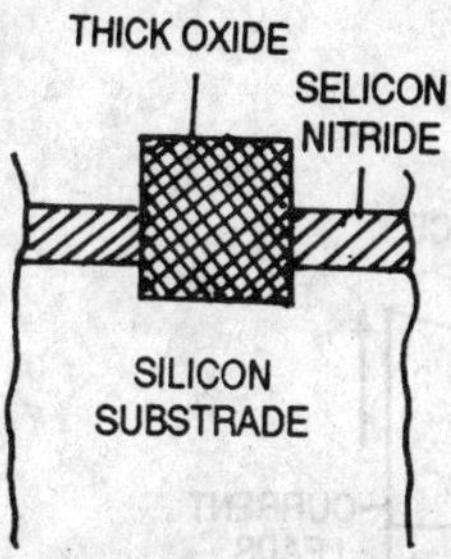

(a) Coplanar process with nitride etch

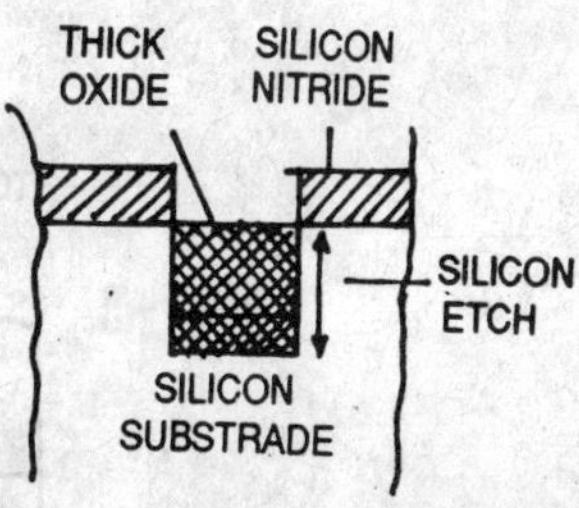

(b) Nitride plus silicon etch

Core : It is the ferromagnetic portion of the magnetic circuit of an electromagnetic device. A simple *ferrite* core is a solid piece of ferromagnetic material suitably shaped into a cylinder, torroid, etc. It should be noted that a laminated core is composed of laminations of ferromagnetic material insulated from each other; eddy currents are thus reduced. It should be noted that a wound core is one constructed from strips of ferromagnetic material wound spirally in layers.

It is *syn.* to core store. It is defined as a static nonvolatile random access memory that consists of an array of rigs of ferrite material strung on a grid of wires.

Core Loss : It is *syn.* to iron loss. This is total energy dissipation in the ferromagnetic core of an inductor of transformer. The energy loss is mainly due to eddy currents and hysteresis loss in the core.

Core-Type Transformer : It is a transformer in which most of the core is enclosed by the windings. The core is made up of laminations; generally the yoke is built up from a stake of laminations and the windings are formed around this.

Cotton Balance : It is also called electromagnetic balance. It is defined as an absolute means of measuring magnetic flux density B in air. It is capable of extremely high accuracy. This method can only be used for fairly strong fields that are uniform over a reasonable volume.

A long rectangular coil is suspended from an analytical balance with the lower end of the coil suspended in the magnetic flux which is to be measured (*see* diagram).

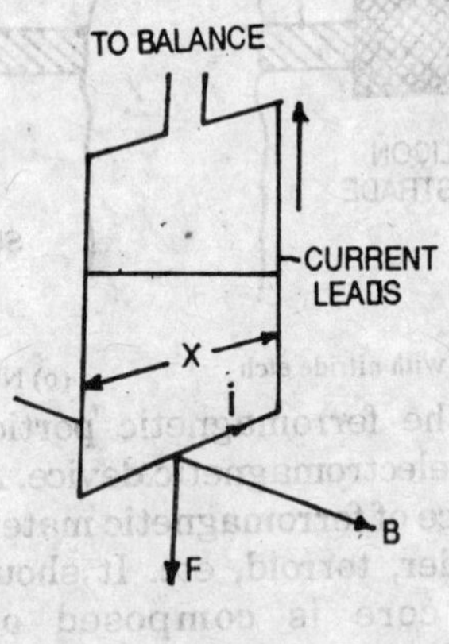

Cotton Balance

Coulobmb Symbol : C This is the SI unit of electric charge, defined as the charged transported through any cross section of a conductor in one second by a constant current of one ampere. Charge, Q, can then be given by equation

$$Q = \int I dt$$

where I is the current.

Coulombmeter : It is also called coulometer. It is defined as an instrument that measures electric charge by the amount of material deposited electrochemically.

Coulomb's Law : It states that the mutual force, F, between two electrostatic point charges, q_1 and q_2, that results from the interaction of the electrostatic fields surrounding them is given by relation

$$F = q_1 q_2 / 4\pi\varepsilon r^2$$

where r is the distance between the charges and ε the permittivity of the medium. Coulomb's law thus relates electrical and mechanical phenomena.

Counter. : It is a device that detects and counts individual particles and photons. The term is used for the detector and the instrument itself.

It is *syn.* to digital counter which is any electronic circuit that counts electronic pulses.

In both cases, the average rate of occurrence of events counted is called the count rate.

Counter Lag : It is defined as the delay between the primary event and start of count.

Counter/Frequency Meter : It is an instrument that may be used as a counter or frequency meter by counting the number of events in a given time.

Coupling : It is the interaction between two circuits in such a way energy is transferred from one to the other. In *common impedance coupling* there is an impedance common to both circuits (Fig. a). It should be noted that the impedance may be a capacitance (*capacitive coupling*), a capacitance and a resistance (*resistance-capacitance coupling*), an inductance (*inductive coupling*) or a resistance (*direct coupling*). The impedance may be a part of each circuit or connected between the circuits. In *mutual-inductance coupling* the circuits are coupled by the mutual inductance, *M*, between the coils L1 and L2 (Fig. b). The coils used are mostly those of a transformer. The use of two separate coils between amplifier stages rather than a transformer is known as chock coupling.

Mixed coupling is defined as a combination of mutual-inductance coupling and common-impedance coupling.

Cross coupling is undesired coupling between communication channels, circuits, or components, particularly those with a common power supply. The removal of unwanted signals, especially those due to cross coupling, is known as decoupling.

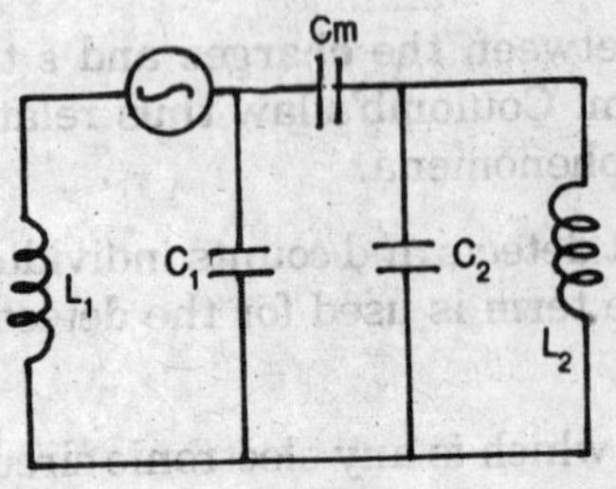

(a) Capacitive coupling

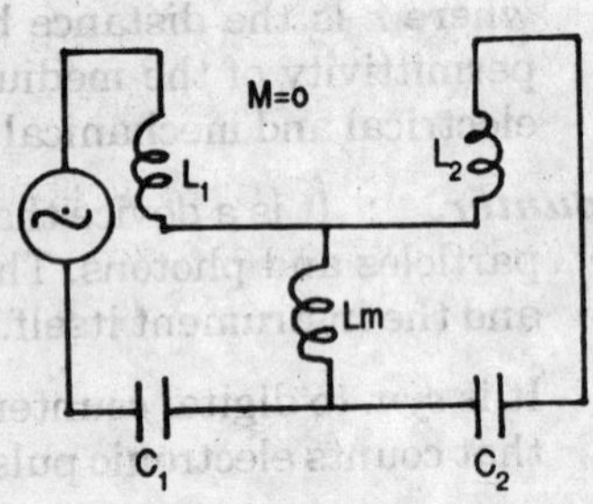

(b) Inductive coupling

Coupling may also be obtained with the help of an electron beam in one circuit in order to influence a second circuit. Such *electron coupling* is generally unidirectional, the second circuit having little influence on the first. It is

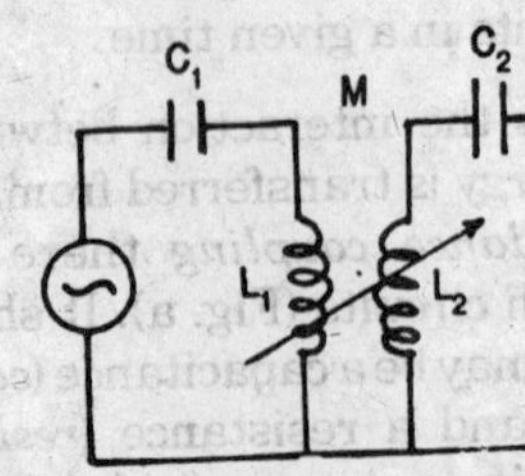

(b) Mutual-inductive coupling

utilized in electron-coupled oscillators, where the load has little effect on the frequency of oscillation.

Crookes dark Space : It is *syn.* to cathode dark space; Hittorf dark space.

Crookes Tube : It is a low-pressure discharge tube used for studying cathode rays.

Crossbar Exchange : It is defined as an automatic telephone exchange in which the necessary connection between telephone lines and exchanges are established by means of a series of crossbar switches so as to establish a desired telephone link.

The crossbar switch has a number of horizontal selector bars and a number of vertical holding bars to which the telephone lines are connected. Contact is made between lines at the crossing points. Contact fingers on the appropriate sector bar are brought into contact with each of the vertical holding bars and the required contact is formed by means of a small electromagnet on the appropriate holding bar which traps the contact finger. The selector bar then comes back to its original position.

Crossover : It is a point at which two conductor, insulated from each other, cross paths.

Crossover Network : It is *syn.* to dividing network. It is defined as a type of filter circuit that divides the frequency range passed between two paths. Frequencies above a specified value pass through one path and those below that value through another.

Crossover Frequency : This is defined as the frequency at which the output passes over from one channel to other channel. The frequency at output is equal.

Crosstalk. : It is the interference due to cross coupling between adjacent circuits or to intermodulation of two or more carrier channels, producing an unwanted signal in one circuit when a signal is present in the other.

Cryotron : It is a very low temperature switching device, depending on superconductivity. It may either consist of a superconductive wire surrounded by a coil or a thin-film form of two insulated crossing strips made of superconductive material with different critical field curves, lead and tin are examples of this.

Crystal-Controlled Transmitter : It is defined as a transmitter in which the carrier frequency is produced by a piezoelectric oscillator.

Crystal Counter : It is a radiation counter that detects and counts subatomic particles. A particle or photon striking a crystal with a potential difference across it produces electron-ion pairs that increase its conductivity. The current pulses produced may then be counted.

Cryctal Cutter : It a device of cutting gramophone records using a piezoelectric crystal. The electrical signals from the recording system cause mechanical displacements in a piezoelectric crystal used as a cutting stylus.

Crystal Filter : It is a filter using piezoelectric crystals to provide its resonant or antiresonant circuits.

Crystal Loudspeaker : It is defined as a loudspeaker in which the sound waves are produced by the mechanical vibrations of a piezoelectrical crystal.

Crystal Microphone : It is a type of microphone containing a piezoelectric crystal in the form of two plates separated by an air gap.

Curie Point : It is *syn.* to magnetic transition temperature; ferromagnetic Curie temperature.

Current : Symbol: I; unit: ampere. It is defined as the rate of flow of electricity. A conduction current is a current flowing in a conductor due to the movement of electrons or ions through the material, generally under the influence of an applied field. The net current is the algebraic sum of the charges.

A *unidirectional current* is one that always flows in the same *direction in a circuit.*

Current Density : It is represented by symbol j or J is defined as a vector quantity that is equal to the ratio of the current to the cross sectional area of the current-carrying medium. The medium may either be a radiation beam or a conductor. The current density is defined either at a point or as the mean current density.

Cut-off : The point at which the current flowing through an electronic device is cut off by the control electrode is called out of point. In a transistor, the cut-off point is the minimum base current at which the device conducts; in a valve it is the minimum negative grid voltage required to stop the current.

Cut-Out Frequency : It is defined as the frequency at which the attenuation of a passive network varies from a small value to a much higher value; this is the theoretical cut-off frequency.

Cut-Off : It is a switch, especially a protective device that is operated, usually automatically, under fault conditions.

Cycle : It is an orderly set of changes regularly repeated. It is also one complete set of changes in the values of a recurring variable quantity, such as an alternating current.

Cyclotron : It is defined as an accelerator that produces a beam of high-energy particles. The particles travel within to semicircular hollow metal electrodes. A unipolar magnetic field-H, at right angles to the plane of the dees, results in the particles to execute circular orbits within the dees, the orbital radius being proportional to particle velocity and magnetic field. Particles are accelerated by a radiofrequency field when they pass between the dees and the orbital radius thus increases. The transit time, t, within a dee is given by the equation

$$t = \pi m / He$$

for a particle of mass m and charge e, and is constant until the relativistic mass change causes it to become out of synchronization with the accelerating field.

Cylindrical Winding : It is a type of winding in which a coil is helically wound, either as a single layer or in multiple layers.

D

Damped : It denotes the progressive diminution of a free oscillation due to an expenditure of energy.

Damping : It is a term used to mean both the cause of the energy loss-friction, eddy currents, etc., and the progressive decrease in amplitude. The amount of damping is termed *critical* if the system just fails to ocillate *(see* diagram). Greater or lesser degrees of damping lead to *overdamping* and *underdamping* respectively.

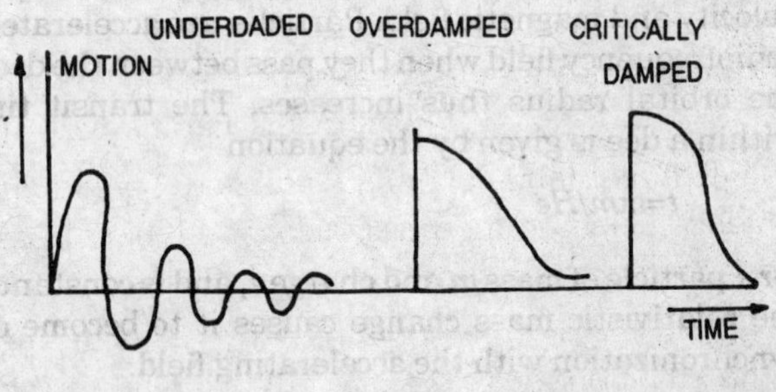

Damping

Damping Factor : Of underdamped oscillations it is the ratio of the amplitude of any one of the damped oscillations to that of the following one.

Daniell Cell : It is a cell in which two different electrolytes are separated by a porous partition. The positive electrode is copper immersed in copper sulphate solution; the negative one is zinc amalgam immersed in dilute sulphuric acid or sometimes zinc sulphate solution.

Daraf : Its symbol: is F^1. The reciprocal farad, used as a unit to measure the reciprocal of capacitance, I.e., elastance.

Dark Conduction : It is defined as low-level conduction that occurs in a photosensitive material when it is not illuminated. If such material is used as a photocathode in devices such as photocells, it gives rise to a dark current.

Darlington Pair : It is defined as a compound connection of two transistors that operates as if it were a single transistor with an extremely high forward-current transfer ratio. The input signal is placed at to the base of transistor T_1 and the emitter current supplies the input to transistor T_2.

D'Arsonval Galvanometer : It is a direct-current galvanometer in which the current to be measured passes through a small rectangular coil suspended between the poles of a permanent horseshoe magnet. The magnetic field given out pay the coil reacts with the field of the magnet forming a torque and causing the coil to rotate about the vertical axis in the field. D'Arsonval movement is utilized in most forms of galvanometer since it combines a high degree of sensitivity with low resistance and high damping.

Dashpot : It is a means of providing mechanical damping in an electrical device. It consists of piston in a cylinder containing fluid. The friction between the piston and the fluid provides the damping.

Data Processing : The automatic or semiautomatic organization of numerical data in a desired manner is called data processing.

D.C. : It is the abbrev. for direct current.

D.C. Amplifier : It is the *abbrve.* for direct coupled amplifier.

D.C. Voltage : This is a informal term for *direct voltage* i.e., a unidirectional voltage that is substantially constant.

Dead Time : It is defined as the time interval immediately following a stimulus during which an electrical device does not respond to another stimulus.

De Broglie Waves : A set of waves that are associated with a moving particle and represent its behaviour in certain situation, as when a beam of particles undergoes diffraction. The wavelength is given by the equation

$$\lambda = h/mv$$

where h is the Planck constant and m and v the mass and velocity of the particle.

Debunching : It is defined as the spreading of electrons in an electron beam or in a velocity modulated tube that results from their mutual repulsion. The angle of spread of the electron beam is the divergence angle.

Deca : Its symbol is da prefix to a unit, denoting a multiple of 10 of that unit.

Deci : Its symbol is d. It is a prefix to a unit, denoting a sub-multiple of 10^{-1} of that unit.

Decibel : It is a convenient arrangement for expressing amplification, attenuation, gain or loss in small, workable numbers. It is ten times the common logarithm of the power ratio. Thus if two values of power, P_1 and P_2 differ by n desibels then

$$n = 10 \log_{10} (P_2/P_1)$$

$$\text{i.e., } P_2/P_1 = 10^{n/10}$$

If P_1 and P_2 are the input and output powers, respectively, of an electric network then if n is positive, i.e., P_2P_1, there is a gain in power; if n is negative there is a power loss.

The *bel*, symbol : B, is equal to 10 decibels. Since it is inconveniently large the decibel is used in practice. If two power values differ by N bels then

$$N = \log_{10} (P_2/P_1)$$

It is customary to abbrivate decibels as db.

Defect Conduction : Conduction in a semiconductor due to the presence of holes in the valence band. The presence of the holes is due to an imperfection in the crystal lattice of the semiconductor.

Deformation Potential : It is defined as an electric potential caused by mechanical deformation of the crystal lattice of semiconductors and conductors.

Degeneracy : It is a condition that arises when an atomic or molecular system with a number of possible quantized states has two or more distinct states of the same energy. The number of degenerate states each possessing that energy level is called the *statistical weight*.

Degenerate Semiconductor : It is a semiconductor that has the Fermi level located either in the valence or in the conduction band, giving rise to essentially metallic properties over a wide range of temperature.

Degree Celsius : It is *syn.* to degree centigrade. Its symbol is C.

Dekatron : It is a type of multielectrode cold-cathode scaling tube that has ten sets of electrodes that function in turn. As voltage pulses are received, a glow discharge moves from one set of electrodes to the next. A visual display of counts in the decimal system is provided using the tubes in cascade.

Delay : It is defined as the time interval between the propagation of a signal and its reception; the time taken for a pulse to traverse any electronic device or circuit.

Delayed Automatic Gain Control : It is *syn.* for biased automatic gain control.

Delay Equalizer: : It is defined as a network or filter that compensates for the effects of delay distortion and thus maintains the wavefront of a transmitted wave.

Delay Line : It is any circuit, device, or transmission line that introduces a known delay in the transmission of a signal. Coaxial cable or suitable L-C network may be used to provide short delay times but the attenuation is usually too great when longer delay times are required.

Depletion Layer : It is defined as space charge region in a semiconductor which has a net charge due to insufficient mobile charge carriers. Depletion layers are mostly formed at the interface between two dissimilar conductivity types of semiconductor, in the absence of an applied voltage. A depletion layer can also be obtained at the surface of a semiconductor of given conductivity type under the influence of an electric field.

Depletion Mode : It is a means of operating field-effect transistors in which increasing the magnitude of the gate bias decreases the current.

Deposition : It is the application of a material to a base by means of vacuum, electrical, chemical, screening, or vapour techniques.

Derating : It is the reducing the maximum performance ratings of electronic equipment or devices when operated under uncommon or extreme conditions. This ensures an adequate safety margin.

De Sauty Bridge : It is a four-arm bridge used for the direct comparison of capacitances (see diagram). The capacitors are charged or discharged using the key K; in case no response is observed from the ballistic galvanometer G, then

$$R_1C_1=R_2C_2$$

Detector : It is *syn.* to demodulator. It is defined as a circuit, apparatus, or circuit element that is used in communications to demodulate the received signal, i.e., to extract the signal from a carrier with minimum distortion.

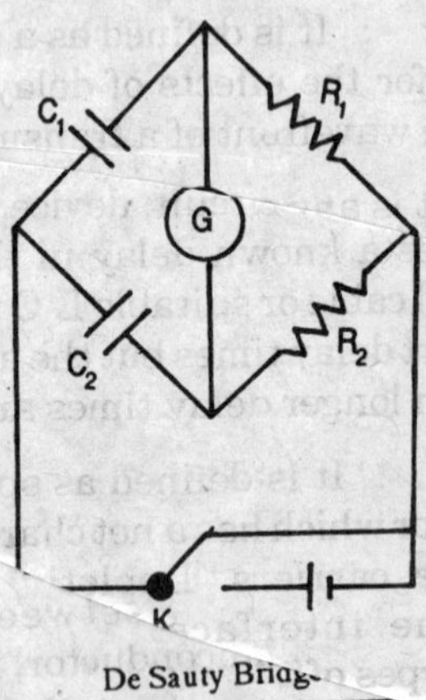

De Sauty Bridge

This term is also used for any device to detect the presence of a physical property or phenomenon, such as radiation.

Detune : It is a term used to adjust the frequency of a tuned circuit so that it differs from the frequency of the applied signal.

Deviation : The difference between the observed value of a measurement and the true value is called deviation. In automatic control systems it is the difference between the instantaneous value and the desired value.

Device : It is an electronic part that contains one or more active elements, such as a transistor, diode, or integrated circuit.

Diacritical Current : It is the current required in a coil to produce a flux equal to one half of the saturation flux.

Diamagnetism : It is defined as a very weak effect that is common to all substances and occurs as a result of the orbital motion of electrons around the nucleus of atoms. It should be kept in mind that diamagnetism is independent of the temperature of the material.

When a substance is placed in a magnetic flux density B, each individual electron experiences a force due to B as it is a moving charge. The orbits and velocities of the electrons are changed in order to produce a magnetic flux density that opposes B. Each orbital electron thus acquires and induced magnetic moment that is proportional to B and is the opposite sense; thus the sample has a negative magnetic susceptibility.

Diaphragm : It may be defined in the following manners:

1. A vibrating membrane used in sound transducers, like microphones, loudspeakers, and the humand ear.

A porous partition in an electrolytic cell which separates the electrolytes while permiting ions to pass through.

Dichroic Mirror : It is the mirror used in a colour television camera to reflect one designated visible frequency band and transmit the remaining.

Dicke's Radiometer : It is a instrument which measures microwave noise power precisely by comparing it with the noise from a standard source in a waveguide.

Dielectric : It refers to solid, liquid, or gaseous material which can sustain an electric field and act as an insulator.

Dielectric Heating : It is a method of heating an insulating material, by a high-frequency alternating electric field.

Dielectric Isolation : It is a process of isolating individual regions in an integrated circuit, specifically a bipolar integrated circuit, by surrounding the region with insulating material (dielectric) rather than with isolating diffusions.

Dielectric Phase Angle : It means the difference between the phase angle of the alternating (sinusoidal) voltage applied across a dielectric material and the phase angle of the resulting alternating current. The difference between the dielectric phase angle and 90° is the called *dielectric loss angle*. It should be noted that the cosine of the dielectric phase angle (or the sine of the dielectric loss angle) is called the *dielectric power factor*.

Dielectric polarization or Electric Polarization : Symbol: P; unit: coulomb per metre squared. An effect observed in a dielectric when electric field is applied.

Dielectric strength or Disruptive Strength. : Unit: volts per metre. The maximum electric field that can be sustained by a dielectric before breakdown takes place.

Differential Amplifier : An amplifier having two inputs and which produces an output signal that is a function of the difference between the inputs.

Differential Analyser : It refers to a device for solving differential equations.

Differential Capacitor : It is a variable capacitor having two sets of fixed plates and one set of moving plates. As the moving plates rotate between the fixed plates, the capacitance to one of set of plates is increased and that to the other is decreased.

Differential Galvanometer : It is a type of galvanometer which gives a deflection that is a function of the difference between two currents.

Differential Resistance : The resistance of a device or component part measured when small signal conditions are used.

Differential Winding : Two or more coils or two windings of a single coil arranged in such a way that when carrying a current their magnetomotive forces are in opposition.

Differentiator : A differented is defined as a resistance capacitance circuit (Rc) having a short time constant. Some input or pulse train is applied across the circuit and an output is taken across the resister.

Diffraction : A phenomenon occurring when electromagnetic waves or beams of charged particles, like electrons, encounter either an opaque object or a boundary between two media. It should be noted that the beams are not propagated strictly in straight lines but are bent at the discontinuity. This effect is due to the wave nature of electromagnetic radiation.

Diffused Junction : A junction between two different conductivity regions within a semiconductor obtained by diffusion of the appropriate impurity atoms into the material.

Diffusion : It is the process of introducing selected impurity atoms into designated areas of a semiconductor in order to modify the properties of that area. First of all the semiconductor is heated to a predetermined temperature in a gaseous atmosphere of the desired impurity. Now impurity atoms condense on the surface diffuse into the semiconductor material in both the vertical and horizontal directions. The numbers of impurity atoms and distance travelled at any given temperature is given by Fick's law. The interface between two different conductivity regions within a semiconductor is a diffused junction.

Double diffusion is a method of obtaining diffused junctions in which successive diffusions of different impurity types are made into the same well-defined regions of semiconductor.

Digital circuit : It is circuit which responds to discrete values of input voltage and produces discrete values of output voltage.

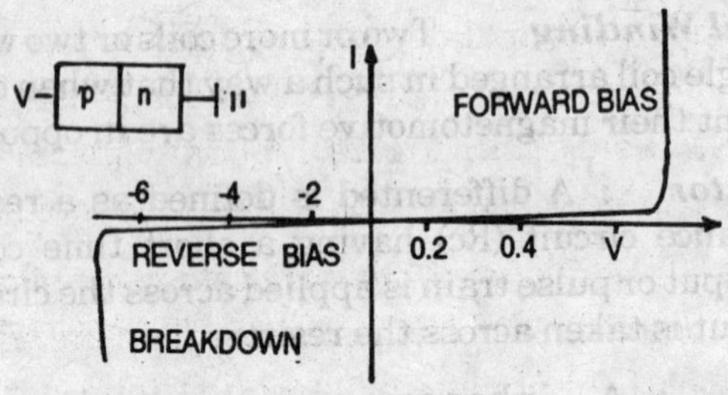

I-T characteristic for semicondutor diode

Digital Voltmeter. (DVM) : It is a voltmeter which displays the values of the voltage as numbers. The input is generally given as an analog signal and this is repetitively sampled by the voltmeter, which displays the instantaneous values.

Digitron or Nixic Tube : A type of cold cathode scaling tube which has several cathodes (usually ten) shaped into the form of characters (usually the digits 0 to 9).

Diode : It is two electrode vacuum tube consisting of a cathode that emits electrons by thermionic emission surrounded by a plane. There are several different types of diode, their voltage characteristics determining their application. Diodes are used as rectifies; those used for other purposes have special names, as with the Gunn diode.

A *semiconductor diode* is made of a simple *p-n* junction. Current flows when a forward bias is applied to the diode and increases exponentially (see diagram).

Diode Forward Voltage : The voltage across the electrodes of a diode when current flows is known as the diode forward.

Diode Transitor Logic : (DTL) It consists of a family of integrated logic circuits in which each input signal comes through a diode and the output is taken from the collector of an inverting transistor. The basis circuit in this case is a NAND gate.

Diplexer : An instrument which permits the simultaneous transmission of two signals in the same circuit of channel without interaction.

Dipole : It means a system of two equal and opposite charges very close together. A uniform field gives a torque which aligns the dipole along the field without translation. The product of one of the charge and the distance between them is the *dipole moment* (symbol : P). Dipole moment is related to the electric field strength, E, and the torque, T, by the relation:

$$T = p \times E$$

Dipole Aerial : It is an aerial commonly used for frequencies below 30 megahertz.

Direct-Coupled Amplifier : *Abbrev.* d.c. amplifier. It is an amplifier having the output of one stage fed directly to the input of the next or through a resistance. This type of an amplifier is used for direct-current amplification.

Direct-Coupling or Resistance Coupling : Coupling between electronic circuits or devices which is not frequency dependent.

Direct Current (d.c.) : A unidirectional current of substantially constant value.

Direct-Current Restorer : It is a device which restores the d.c. component or low-frequency component to a signal which has had its low - frequency components removed by a circuit element with high impedance to direct current. The device may add d.c. or low frequency to a signal lacking these components.

It should be mentioned here that direct current restoration is used in television receivers to reconstruct the original video signal.

Direct-Current Transmission : It is a process of transmission used in television in which the direct-current component of the luminance signal is directly shown by the transmitted signal.

Direct-Gap Semiconductor : Solution of Schrodinger's equation for an electron moving through a periodic potential which changes with the spacing of atoms is a crystal lattice only exist for certain values of energy, which give the permitted energy bands; other values-forbidden bands-cannot exist.

Directive Aerial or Directional Aerial : An aerial which is more effective transmitter or receiver of energy in certain directions than in others. A common method of obtaining such directivity is by using a *passive aerial* and an *active aerial* in conjunction.

Direct Ray : It refers to that path along which a radiowave travels in the minimum possible time between a transmitting and receiving aerial.

Direct Wave : It refers to that portion of a transmitted wave which travels along the path of a direct ray.

Disc Colloq : It is a gramophone record.

Discharge : It can be defined in the following ways to remove or reduce the electric charge on a body like a capacitor.

(i) The passage of an electric current of charge through a medium, when accompanied by luminous effects.

(ii) The conversion of chemical energy into electricity in a cell by drawing current from it.

Discharge tube Rectifier : It is a gas-discharge tube which may be used as a rectifier, the electrodes being arranged in such a way that the discharge current in one direction is greater than that in the opposite direction upon reversal of the applied voltage.

Discriminator : It refers to a circuit that converts a frequency-modulated or phase-modulated signal into an amplitude-modulated signal.

It also refers to a circuit that selects signals with a particular range or amplitude or frequency and rejects all others.

Disc Winding : A type of winding for transformers having flat coils in the form of a disc. Disc windings are mostly employed for high voltage applications.

Disintegration Voltage : The voltage at which the cathode of a gasfilled tube starts to disintegrate or a result of bombardment by positive ions. The voltage at which cathode sputtering starts.

Displacement : Or electric displacement; dielectric strain; electric flux density. Symbol D; unit: coulombs per metre squared. A vector quantity defined by the equation.

$$D = \epsilon_0 E + P$$

where E is the electric field strength, P the dielectric polarization and ϵ_0 the permitivity of free space.

Displacemant Current : It is the rate of change of electric flux in a dielectric with respect to time ($\partial D/\partial t$) if the applied electric field changes.

Display : If refers to the visual presentation of information as on the screen on a cathode-ray tube or by using scaling tubes in cascade, as we do in case of digital voltmeter.

Disruptive Discharge : It is regarded as a sudden large increase in current in an insulator, semiconductor, or gas in or case the material breaks down under the influence of an applied electric field.

Dissipation or Loss : It means loss of power due to the tendency of electronic circuits and components to resist the flow of current.

Distortion : Any change in the wave form of a signal as it passes through an amplifier or other electrical circuit in called *distortion*. It has to be noted that a charge in amplitude can occur without wave form being altered and is not distortion. Different types of distortions are as follows

Amplitude distortion takes place if the ratio of the rootmeonsquare value of the output to the r.m.s. value of the input varies with the amplitude of the input, the condition is that both waveforms are sinusoidal. In case harmonics are present in the output waveform only the fundamental frequency is considered.

Attenuation distortion takes place when the gain or

loss of the system depends on frequency.

Grossover Distortion takes place in push-pull operation when the transistors are not operating in the correct phase with each other.

Delay distortion is regarded as a change in the waveform because of the variation of the delay with frequency.

Deviation distortion takes place in frequency-modulated receivers which have an inadequate handwidth or nonlinear discriminator.

Harmonic distortion is because of harmonics not present in the original waveform.

Intermodulation distortion takes place from spurious combination frequency components in the output of a nonlinear transmission system when two or more sinusoidal voltages, applied simultaneously form the input.

Nonlinear distortion takes place in a system when the instaneous transmission properties depend on the magnitude of the input.

Phase distortion takes place when the phase change introduced is not a linear function of frequency.

Optical distortion is regarded as an image when seen in electronic systems, such as cathode-ray tubes television picture tubes, etc., and in facsimile transmission. It is due to errors in the electron-lens focusing systems.

Aperture distortion of an image takes place in a scanning system if the scanning spot has finite dimensions rather than infinitely small dimensions.

Barrel and pincushion distortion are seen when the lateral magnification is not constant rather depend on image size (see diagram).

Coma distortion when the focusing elements of the electron gun are misaligned.

Keystone distortion is as a result of the length of the horizontal scan line varying with the vertical displacement of the line.

Origin distortion is a flattening of the waveform at the

place it crosses the zero line in a gas-focused cathode-ray tube employing electrostatic deflection. It takes place at low deflecting voltages when there is a nonlinear relationship between the angular deflection and the deflecting voltage.

Trapezium distortion is a trapezoidal pattern on the screen of a cathode-ray tube.

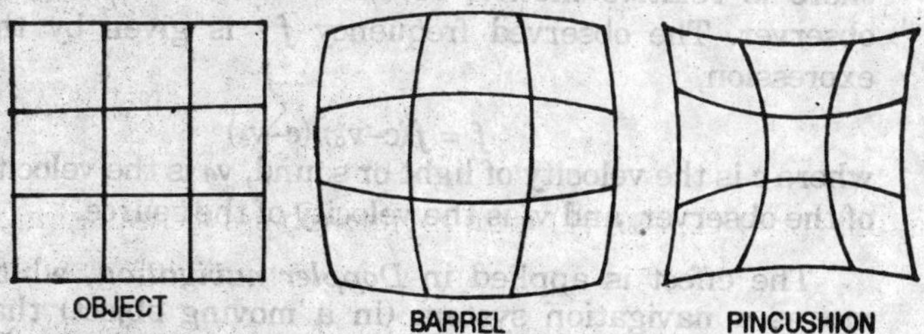

Optical distortion

Distributed Capacitance : It is defined in the following two ways.

(1) the capacitance of an electrical system, like a transmission line, considered to be distributed along its length.

(2) The capacitance between individual turns on a coil or between adjacent conductors.

Distributed Inductance : The inductance of an electrical system, like transmission line, considered to be distributed along its length.

Diversity System : A communication system having two or more paths or channels.

Divider : A circuit which reduces the number of pulses or cycles by an integral factor.

D/MOS : The circuits or transistors which are fabricated using double diffusion.

Doping : It refers to the addition of a particular type of impurity to a semiconductor in order to get a desired n-conductivity or p-conductivity.

Doping Compensation : If a particular type of impurity to a semiconductor is added in order to compensate for the effect of an impurity already present, this is called doping compensation.

Doping Level : It is the amount of doping necessary to achieve the desired characteristic in a semiconductor.

Doppler Effect : The change in the apparent frequency of a source of electromagnetic radiation (or sound) in case there is relative motion between the source and the observer. The observed frequency f' is given by the expression,

$$f' = f(c - v_o)/(c - v_s)$$

where c is the velocity of light or sound, v_0 is the velocity of the observer, and v_s is the velocity of the source.

The effect is applied in *Doppler navigation*, which means a navigation system (in a moving object) that operates by ground reflection. *Doppler radar* uses the Doppler effect to distinguish between fixed and moving targets: the measurement of the change in the frequency of the reflected wave is used to determine the velocity and direction of the moving target.

Dot Generator : A test generator which is used with a television receiver to adjust the convergence of the picture tube.

Double Current System : The telegraph system which reverse the direction of the electric current in order to effect transmission of the signals.

Double Modulation : It is multiple modulation involving two carriers only.

Double Pole Switch : It is the switch which can operate simultaneously in two independent circuits.

Double Sideband Transmission : Transmission of both sidsbands produced in case a carrier wave is amplitude modulated, but not of the carrier itself.

Drain : It is regarded as the electrode of a field-effect transistor through which carriers leave the interelectrode space.

Drift : The variation with time of any electrical property of a circuit or apparatus. It mostly occurs during warm up or when the device is nearing the end of its useful life.

Drift Mobility of Carrier Mobility : Symbol: μ The average velocity of excess minority carriers in a semiconductor per unit electric field.

Drift Rate : See drift.

Drift Space : It is the region in an electron tube which is free of electric or magnetic fields.

Drift Transistor or Graded-Base Transistor. : A transistor having the impurity concentration in the base changes smoothly across the base region. It should be noted that the doping level is high at the emitter base junction dropping to a low doping level at the base collector junction. The impurity profile and associated energy levels of an *n-p-n* transistor are illustrated in the diagram.

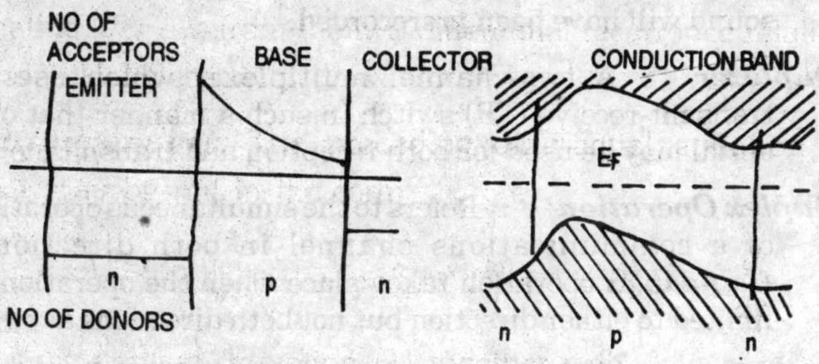

Variation of doping levels and energy band in drift transistor

Drift Velocity : It is the average velocity of electrons in a conductor or of ions in a gas and in the direction of applied electric field.

Driver : It refers to a circuit or device which provides the input for another circuit or controls the operation of applied electric field.

Driving-Point Impedance : It is regarded as the ratio at the input terminals of a network of the root-mean square (rms) value of the applied sinusoidal voltage to the rms value of the resulting current between the terminals.

Drop-in : It means the presence of a spurious bit in the information recorded at the input of a digital computer due to faulty reading of the input/output signals.

Dropping Resistor or Dropper. : It is a resistor which is put in a circuit to provide a voltage drop across its terminals and thus reduce the voltage in the circuit.

Dry Joint : It is a faulty soldered joint having a high resistance due to a residual oxide film.

Dual-Channel Sound : A method used in television receivers to separate the sound and video signals after the common first detector stage. It should be kept in mind that the separate intermediate-frequency stages are employed for each signal.

Dual in Line Package : It is a standard form of package used for integrated circuits.

Dubbing : The combining of two sound signals into a composite recording. In this case minimum one source of sound will have been prerecorded.

Duplexer : A two-channel multiplexer which uses a transmit-receive (TR) switch in such a manner that one aerial may be used for both reception and transmission.

Duplex Operation : Refers to the simultaneous operation of a communications channel in both directions. *Half-duplex* operation takes place when the operation is limited to either direction but not both directions at once.

Dust Core : It is a magnetic core which is made from a powdered material like as ferrite. These cores have a very low eddy-current loss at high frequencies.

Duty : It is regarded as the statement of the operating conditions and their durations to which a device is subjected, including rest and de-energized periods.

Dynamic : It denotes any electrical device, circuit, or apparatus in which the electrical parameters are constantly changing. The term can be applied to those devices which operate with alternating currents and voltages, especially those with a marked frequency dependence.

Dynamic Impedance or Dynamic Resistance : The imp- edance at resonance of a parallel resonant circuit.

Dynamic Memory : It means a solid-state memory whose stored information decays over a period of time. The decay time can range from milliseconds to seconds depending on the nature of the device and its physical environment. The memory cells must undergo refresh operations periodically to maintain the integrity of the stored information. It should be noted that MOS random-access memory and CCD memory are both dynamic memories.

Dynamic Range : The range in which an active electronic device can give a suitable output signal in response to an input signal. It is said to be the difference in decibels between the noise level of the system and the level at which the output is saturated.

Dynamic Resistance : It is regarded as the resistance of any electronic device under normal operating conditions.

Dynamo : An electromagnetic generator which gives either alternating or direct current.

Dyne : It is the unit of force in the CGS system of units. If we convert into S.I. units then one dyne equals 10^{-5} newton.

Dynode : This is a small plate constructed of a material containing many loosely held electrons.

E

Earphone : It is regarded as a small loudspeaker that is designed to be used very close to the ear. Its main uses include as hearing aids, the receiver in a telephone, a.c. bridge measurements and use with reproduction systems such as radio. Two earphones used together give a *headset.*

Earth : U.S. *syn.* ground. (i) A large conducting body, like the earth, when is taken to be the arbitrary zero in the scale of electrical potential.

(ii) A connection, which may be accidental, between a conductor and the earth.

Earth Capacitance : It refers to the capacitance between any circuit or equipment and a point at earth potential.

Earth Current : (i) A current which flows to earth, especially one that result from a fault in a system.

(ii) Any current flowing in the earth. Mainly earth currents are associated with ionospheric disturbances. Buried cables sometimes have their lead sheaths corroded by a direct earth current.

Earth Fault : The fault which takes place when a conductor is accidentally connected to earth or in case the resistance to earth of an insulator falls below a specified value.

Earth potential or Zero Potential : The potential of a large conducting body, like the earth, taken to be the arbitrary zero in the scale of electrical potential.

Earth-Return Circuit : It is a circuit which consists of one or more conductors in parallel and connects two points in a telecommunication system and that is completed through earth at the two points.

Echo : (i) A communications wave which has been reflected or refracted and has sufficient magnitude and delay to distinguish it from the direct wave. In radio an echo is heard. In television it appears as a ghost on the screen; it may or may not be heard simultaneously.

(ii) The portion of a transmitted radar signal reflected back to the receiver.

Echo Sounding : A system based on the same principles as radar using sound waves instead of radiowaves.

ECH Set of Eddy-Current Heating Set : A radiofrequency generator which is used to produce the alternating magnetic field used for induction heating.

Eddy Current : It is the current induced when a conductor is subject to a varying magnetic field. Eddy currents in a moving conductor react with the magnetic field to give retardation of the motion and are used to produce damping.

Edge Connector : It is on a track on a printed circuit board which is taken to one edge of the board to form a connector. It is to be noted that each board has several edge connectors, which may be plugged into a suitable socket permitting connections to be made to the circuit on the board.

Edge Effect : It is the deviation from parallel in the lines of force representing the electric field at the edges of parallel-plate capacitors. It gives a field that is nonuniform at the edges.

Edison cell : It is a secondary cell in which the plates are impregnated steel grids. The positive plate is a nickel/nickel hydrate mixture and the negative plate is iron oxide. Potassium hydroxide is the electrolyte.

Effective Resistance : It is the resistance to alternating current of a conductor or other circuit element. It is recorded as the power in watts dissipated as heat divided by the current in amperes squared.

e h t. or E.H.T. : It is the abbrev. for extra high tension. Thus term is applied to the high-voltage supply for cathode-ray tubes or television picture tubes.

Einthoven galvanometer or String Galvanometer : A galvanometer containing a tightly strung conducting thread between the poles of a strong electromagnet. A current passed through the thread results in its deflection at right angles to the magnetic field. The deflection is observed with a high-power microscope. The instrument is quite sensitive and can detect current of 10^{-11} ampere.

Elastance : Symbol: S; unit: farad^{-1}. It is the reciprocal of capacitance.

Elastoresistance : A change in the resistance of a meterial in case it is stressed within its elastic limit.

Electret : A substance which is permanently electrified and has oppositely charged extremities: it is the electrical analogue of a permanent magnet. The applications of electrets are in electrometers and in capacitor microphones.

Electric Axis : It means, the direction in a crystal of maximum conductivity. It is the X-axis of a piezoelectric crystal.

Electric Degree : It is defined as 1/360 part of an alternating current cycle. Currents or voltages arising in different parts of a circuit may be written in terms of vectors.

Electric Eye : *Colloq*. A photocell or photovoltaic cell, specifically used in burglar alarm systems, lifts, etc.

Electric Field : The electric field at a point is defined as the force that acts on a unit position change placed at that point

$$E = \frac{F}{q} \; newtons/coulomb$$

when E = electric field

F = force acting or one charge q at the point.

Electric Field Strength : Symbol: E; unit: volts per meter. It is the strength of an electric field at a point, measured in terms of the mechanical force per unit charge. The force, F, given by Coulomb's law, is thus related to field strength by the equation:

$$F = eE$$

where *e* is the electron charge.

Electric Flux Symbol : Unit: Coulomb. It is the quantity of electricity displaced across a given area in a dielectric. It is given by the scalar product *D*.d*s* of the electric displacement, *D*, and an element d*S* of area. Gauss's theorem can be written mathematically as

$$\int D.dS = \int \rho e d\mathrm{r}$$

where p_e is the volume electric charge density in a small volume element dr. It means total flux through an area surrounding a charge *q* is thus equal to *q*. It should be noted that this flux is unaltered by the presence of a dielectric medium.

Electric Heating : It is the production of heat from electric energy. It can be obtained by passing a current through a resistance, use of an electric arc, induction heating, and dielectric heating.

Electric Hysteris Loss : Electrical loss in a dielectric material as a result of internal forces in the material produced by a varying electric field. The loss usually appears in the form of heat.

Electric Image : It is defined as an electrical point or system of point on one side of an electrified surface, which has the same electrical action on the other side of surface, similar to the electrification of the surface. "This concept is designed by kelvin to solve problems of distribution of electric field by method of electrical images.

Electric Interference : It is the interference caused by electrical apparatus and creating disturbances to a signal.

Electricity : The phenomena associated with static or

dynamic electric charges, like electrons.

Electric Potential : symbol: *V*; unit: volt. It is defined as the work required to bring a unit positive charge from infinity to a point in an electric field. The potential is one volt when one joule is required to transfer a charge of one coulomb.

Electric Recording : A sound recording system which employs the amplified signal from one or more suitable microphones.

Electric Screening : It is the surrounding of apparatus with an electrical conductor to prevent interference from undesired electrical disturbances.

Electric Spectrum : The colour spectrum given out by an electric arc.

Electrocapillary Phenomena. : The changes produced in the surface tension of a liquid by an electric field.

Electrocardiograph (e.c.g.). : It is a sensitive instrument which measures and records the voltage and current waveforms produced by the heart muscles of living animals. The trace produced is termed as an electro-cardiogram.

Electrochemical Equivalent : The mass of any ion which is deposited from an electrolyte solution by a current of one ampere flowing for one second.

Electrode : It is a device that emits, collects, or deflects electric charge carriers.

Electrode Dissipation : It is the heat dissipated in a particular electrode, generally the anode, as a result of bombardment by electrons, ions, or radiation from other electrodes.

Electrode Drop. : The potential difference across and electrode caused by its resistance.

Electrode Efficiency. : It is defined as the ratio of the actual mass deposited from solution in an electrolytic cell to the theoretical mass.

Electrodynamic Instrument : It is the instrument in which the interaction of magnetic fields produced by currents in a system of movable and fixed coils gives a torque. These instruments can be operated with both direct and alternating currents.

Electrodynamics : It refers to the study of the mechanical forces acting upon and between currents.

Electrodynamometer : It refers to an electrodynamic instrument which contains two coils, one fixed and one movable, and produces a deflection equal to the square of the current passing through the coils, which are connected in series. It is mostly used as a standard for current or voltage measurements.

When it is to be used as a wattmeter,, the fixed coil is connected in series with the load and the movable coil shunted across a high noninductive resistance (see diagram). It is to be noted that the transformer coupling to the current circuit may be used to isolate the instrument from the main current. The couple produced is then proportional to the wattage.

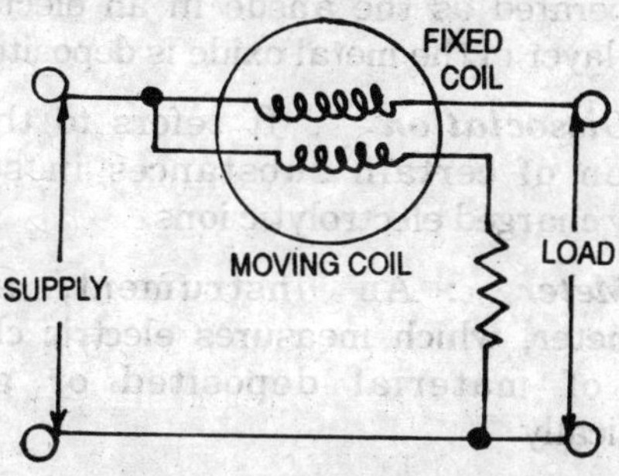

Singel-phase electrodynamic wattmeter

Electroencephalograph (e.e.g.). : It is a sensitive instrument which measures and records the voltage waveforms produced by the brain. The trace produced is termed as an *electroencephalogram.*

Electroluminescence or Destrau Effect : The emission of light by certain phosphorescent substances under the effect of an applied electric field. The phenomenon is employed in the *luminescent lamp*. It consists of a slab of dielectric material within which phosphorescent power is dispersed and which is sandwiched between transparent plate electrodes. Light is produced when voltages of 100 to 500 volts are applied across the plates.

Electrolysis : It is the production of chemical change, generally decomposition, when current flows through an electrolyte.

Electrolyte : A substance which conducts electricity when in solution or when molten because of its dissociation into ions.

Strong electrolytes are compounds, such as material acids, which are completely dissociated into ions when in solution. Weak electrolytes are compounds only partially dissociated in solution. Weak solutions of such compounds are better conductors than strong solutions as a result of their greater degree of dissociation.

Electrolytic Capacitor : Any condenser in which the dielectric layer is formed by an electrolytic method. The capacitor does not essentially contain an electrolyte. When a metal electrode, like an aluminium of tantalum one, is operated as the anode in an electrolytic cell a dielectric layer of the metal oxide is deposited.

Electrolytic Dissociation : It refers to the reversible separation of certain substances in solution into oppositely charged electrolytic ions.

Electrolytic Meter : An instrument, such as a coulombmeter, which measures electric charge by the amount of material deposited or gas evolved electrolytically.

Electrolytic Photocell : An electrolytic cell, made from certain materials such as selenium electrodes in selenium dioxide solution, which is sensitive to light when a small external d.c. voltage is applied across it. The cell has a linear sensitivity of about one milliampere per lumen of luminous flux.

Electrolytic Polarization : A decrease in the potential difference across an electrolytic system caused by polarization in the electrolyte.

Electrolytic Polishing or Electropolishing : Polishing of a metal by making it the anode of an electrolytic cell and passing current through the cell. Under suitable conditions the protuberances dissolve preferentially leaving a smooth shining surface.

Electrolytic Rectifier : A rectifier that has two dissimilar electrodes in an electrolyte. proper combinations of electrodes and electrolyte produce a system that conducts much more easily in one direction than the other.

Electrolytic Separation. : It is the separation of isotopes by electrolysis and is based on the fact that different isotopes of an element are liberated at an electrode at different rates.

Electrolytic Tank : An apparatus that can be used to tell the behaviour of certain physical systems, such as a system of electrodes, by putting a metal scale model of the system in a poorly conducting electrolyte.

Electromagnet : It is a device which is magnetized only when electric current flows in it.

Electromagnetic Deflection or Magnetic Deflection
: It refers to the use of electromagnets to deflect an electron beam. The most important application is in the cathode ray tube in which two pairs of deflection coils produce vertical and horizontal deflections.

Electromagnetic Induction : The production of an electromotive force in a conductor in case there is a change in magnetic flux through the conductor.

Electromagnetic Lens or Magnetic Lens : It consists of an arrangement of coils which focuses an electron beam electromagnetically.

Electromagnetic Radiation : Energy which is radiated by a charged particle in undergoing acceleration.

As moving charged particles have associated wavelike features therefore electromagnetic radiation has a wave/particle duality: it may also be considered as a stream of particles (photons) which move at the velocity of light, *c*, and have zero rest mass. Although wave motion is enough to describe the properties of reflection, refraction, and interference, quantum theory, which is concerned with the particulate nature of electromagnetic waves, must be used to explain phenomena, like the photoelectric effect, that occur when radiation and matter interact.

The properties of the radiation depend on the frequency, v, of the waves. The frequency, wavelength λ and velocity are related by the following equation

$$c = v\lambda$$

The photons have energy E related to the frequency v by the following equation

$$E = hv$$

where h is the Planck constant, Energy is interchanged between radiation and particles by absorption and emission in discrete amounts, termed quanta, the energy of each quantum being hv.

The total range of possible frequencies is called the electromagnetic spectrum. It should be noted that the radiowaves have the lowest frequencies; progressively higher frequencies are associated with infrared radiation, light, ultraviolet radiation, X-rays, through to gamma rays at the highest frequencies.

Electromagnetic Spectrum : It is the whole range of frequency of electromagnetic radiation.

Electromagnetism : It is the magnetic occurrence associated with moving electric charges, the electric phenomena associated with varying magnetic fields, and the magnetic and electric phenomena associated with electromagnetic radiation.

Electromechanical Apparatus : Any apparatus that changes electrical signals into related mechanical movement, or vice versa.

Electrometer : A device which measures potential difference or electric charge by means of the mechanical forces which are supposed to exist between bodies that possess electrostatic charges.

Electromotive Force : (e.m.f. or emf) Symbol; E; unit: volt. The property of a source of electrical energy which results in a current to flow in a circuit. The algebraic sum of the potential difference in a circuit equals the e.m.f., which is measured by the energy given out when unit electric charge passes completely round the circuit. A battery of e.m.f. E will supply a current I to an external resistance R:

$$E = I(R + r)$$

where *r* is the internal resistance of the battry. The term 'electromotive force' strictly applies to a source of electrical energy but many times misused as being equivalent to potential difference.

Electromotive Series or Electropositive Series : If a metal is placed in a molar solution of one of its salts a potential, termed the electrode potential, is developed between the metal and the solution.

The electromotive series gives the relative ability of the elements to form positive ions in solution and in this context it is often referred to as the electrochemical series. Elements higher in the table form positive ions more readily than those lower in the series and will therefore displace lower elements from solution. Thus zinc will displace copper from solution as it is higher in the series. It means to at if a strip of zinc is placed in a solution of copper ions, copper will be deposited and zinc ions will form in the solution.

Electromyograph : It is an instrument, that measures and records the current waveforms generated by contractions in muscles. The trace produced is termed as electromyogram.

Electron : An electron is the lightest of subatomic particles. It carries a tiny negative charge and is said to be surrounded by a minute magnetic field. A stable elementary particle that has a negative charge, *e*, of 1.602 $\times 10^{-19}$ coulomb, mass *m* of 9.100 $\times 10^{-31}$ kg, and spin 1/2. It is the natural unit of electric charge. Electrons are constituents of all atoms, moving around the nucleus in several possible of 'allowed' orbits. Electrons moving in one direction under the influence of an electric field refer to an electric current, the direction of conventional current flow being opposite to the direction of motion of the electrons.

Electron Affinity : (1) Symbol: A or E_a; unit: electronvolt. It is the energy released when an electron becomes attached to an atom or molecule to form a negative ion. Many atoms or molecules have a positive electron affinity, which means the negative ion is more stable than the neutral species.

Electron Beam : A beam of electrons which are mostly emitted from a single souce, such as a thermionic cathode.

Electron-Beam Device : It is a device that utilizes one or more electron beams as an essential part of its operation. It should be noted that several electrodes may be present to form control, and direct the electron beam. The deflected beam is made to strike a fluorescent screen in measuring instruments like the cathode-ray tube.

Electron Density : (i) The number of electrons per unit mass of a material.

(ii) The number of electrons per unit volume of a material.

Electron Emission : It means the liberation of electrons from the surface of a material.

Electron Gas : The concept that the free electrons in a solid, liquid, or gas may be treated as a gas and compared with a real gas dissolved into appropriate material. This type of electron gas obeys a totally different energy distribution than a real gas obeying Fermi-Dirac statistics rather than the Maxwell-Boltzmann statistics obeyed by a ordinary gas. It also has a vanishingly small specific heat.

Electron Gun : A device which produces an electron beam and forms an essential part of many instruments, like cathode-ray tubes, electron microscopes, and linear accelerators, which need electron beams for their operation.

Electronic Device : A device which utilizes the properties of electrons (or ions) moving in a vacuum, gas, or semiconductor.

Electronic Efficiency : It is defined at the ratio of the output power delivered to the load of an oscillator or amplifier at the desired frequency to the average power input to the device.

Electronic Memory : A memory which has no moving parts and in which the read and write operations are entirely electronic.

Electronic News Gathering : It refers to a recording system used in television in which scenes outside the television studio are recorded directly on to videotape instead of on to film. A portable television camera and videotape recorder are used, often in conjunction with a mobile transmitter which relays the recording directly to the main control centre.

Electronics; : It is defined as a field of science and engineering based on effects produced by electrons which move through solid, vacuum and gas.

Electronic Switch : It is an electronic device, like a transistor, that is used as a switch. These devices are mostly operated as high speed switches where a very fast response is required, as for example in a computer.

Electronic Tuning : The use of a coupled electron beam to change the operating frequency of a system. The velocity, intensity, or configuration of the electron beam may be altered to give the desired frequency change.

Electron Lens : A device which is used for focusing an electron beam using either magnetic or electrostatic fields. The two types, which are analogous to optical lenses used with light beams, can be combined in instruments as electron microscopes. Most electron-beam devices, like cathode-ray tubes or camera tubes, use an electron lens to provide a sharp narrow beam.

Electron Microscope : An instrument which uses a beam of electrons to investigate a sample in order to achieve a higher magnification than is possible with an optical microscope.

In case of the *transmission electron microscope* the electron beam is focused by an electromagnetic lens or sometimes an electrostatic lens and has an energy of 50-100 kilovolts (Fig.a). A sharply focused image in one plane can only be obtained with the help of monoenergetic electrons. To avoid energy losses in the beam the sample must be extremely thin (2 nanometres). A resolution of 0.2-0.5 nanometres is possible.

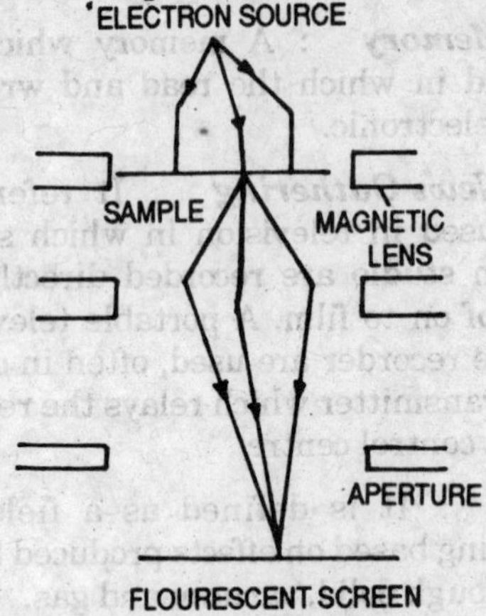

Fig. a. Transmission electron microscope

The *scanning transmission electron microscope* has been made in order to combine the high resolution of the transmission instrument with the three-dimensional image of the scanning type.

Electron Multiplier : An electronic tube in which current amplifications is obtained using secondary emission of electrons. Primary electrons are released from the cathode by some means, like photoelectric effect. These are accelerated by a high potential applied to the first anode. The anodes, termed dynodes, are made from a good secondary emitter and on impact a greater number of

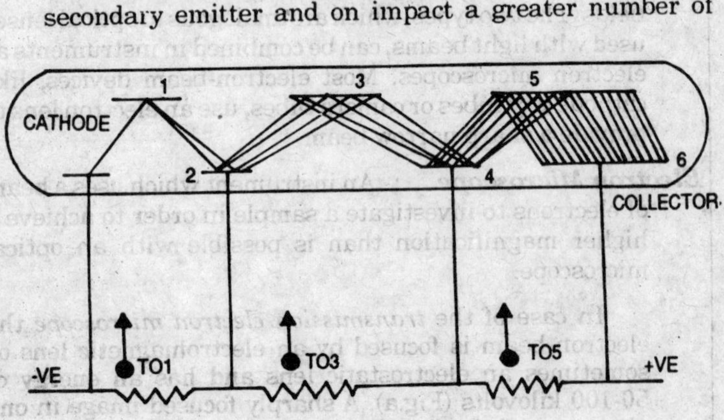

Electron Multiplier

electrons is given out. These are then accelerated in turn with the help of second and subsequent anodes, to each of which an increasing positive potential is applied.

Electron Optics : The study of the behaviour of electron beams under the influence of magnetic and electrostatic fields in a vacuum or very low pressure gas.

Electron Tube : An electronic device in which conduction between two electrodes takes place in an envelope which is sealed or continuously exhausted and has a gas or a vacuum. Tubes which are evacuated to a high vacuum are known as hard tubes; those that contain traces of a gas are soft tubes. The concentration of gaseous atoms in *soft tubes* is less than that of a gasfilled tube but is enough in resulting some modification to the characteristics compared with those of hard tubes. Such tubes frequently contain more than two electrodes.

Electronvolt : Symbol: eV. It is the energy acquired by one electron when passing freely through a potential difference of one volt. The unit is extensively used in atomic physics. One eV equals 1.606×10^{-19} joule.

Electron Voltaic Effect: : It is defined as a phenomenon similar to the photovoltaic effect in which electrons striking the photocathode of a photocell cause electron emission.

Electro-Optics : The investigation of the interactions between the refractive indices of some transparent dielectrics and the electric fields in which they are placed.

Electro-Osmosis or Electroendosmosis : It means the movement of an electrolyte through a fine tube or membrane under the influence of an electric field.

Electrophonic Effect : It refers to the perception of sound when the human body is subjected to alternative currents of particular frequencies and magnitudes.

Electrophonic Instrument or Electrosonic Instrument; Electronic Instrument : A musical instrument which creates sound by electrical or electronic means. There are various kinds of such instruments, the most important of one is the Moog synthesizer.

Electrophoresis : It refers to the movement of colloidal particles in a liquid under the influence of an electric field. It is noticed that the positively charged particles migrate to the cathode (cataphoresis) and negatively charged ones to the anode (anaphoresis).

Electrophorus : It is regarded as an early form of simple electrostatic charge generator.

Electroplating : The application of depositing one metal to the surface of another, for protection or decoration, by electrolysis.

Electropneumatic : It means control systems which contain both electronic and pneumatic elements.

Electropolar : It means magnetic poles or permanent positive and negative charges, as in a magnet or an electret.

Electroscope : An electrostatic instrument that locates small potential differences and electric charges. The most common kind is the *goldleaf electroscope*, which is made of a pair of gold leaves hanging side by side from an insulated metal support enclosed in a draught-proof case. In case a charge is supplied to the support the leaves separate due to their mutual repulsion. A more precise form replaces one of the gold leaves by a rigid metal plate.

Lauritzen's electroscope uses metallized quartz fibre as the sensitive element. The *pith-ball electroscope* utilizes two pithballs suspended by silk threads.

An instrument which is used for accurate quantitative

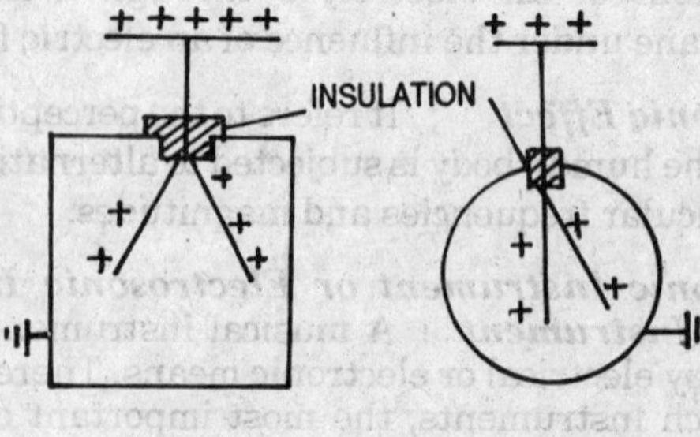

Gold-leaf electroscopes

measurement is called an electrometer.

Electrostatic Adhesion : It is defined as adhesion between two substances or surfaces as a result of the presence of opposite charges, which attract each other.

Electrostatic Deflection : It implies use of electrostatic fields produced between two metal electrodes for deflecting an electron beam.

Electrostatic Field : The electric field associated with charged particles at rest, may be regarded as the region, around a distribution of electrostatic charge, in which a stationary charged particle would experience a force.

Electrostatic Induction : The production of a charge distribution on a conductor when in the nearness of another charged body under the influence of the associated electric field.

Electrostatic Lens. : An electron lens having an arrangement of electrodes that focuses an electron beam electrostatically.

Electrostatic Memory : A type of memory which stores information in the form of electrostatic charges.

Electrostatic Precipitation : A way of precipitating solid or liquid particles form a gas. An electrostatic field is applied across the gas between two electrodes, one of which is earthed. It is to be noted that the particles collect on the earthed electrode.

Electrostatic Printing : A printing in which a pattern of electrostatic charges is formed on the surface of the paper.

Electrostatics : It is the study of electric charges at rest and their associated phenomena.

Electrostatic Separation : It means a method of separating fine powders of different permittivities. The powders undergo various deflections when placed in the intense electrostatic field between two highly charged electrodes.

Electrostriction : It is regarded as change in the dimensions of a body under the effect of an electric field in a medium of relative permittivity different from its own.

Element : An element is a substance which cannot be separated chemically into anything but itself i.e., atom of same atomic number. Elements are the basic substances from which compounds are built up by chemical combination: this obstructs disruptive processes in which the atomic nuclei comprising the element are disturbed.

Elementary Particles : The particles of matter that cannot be sub-divided into smaller particles. These particles are described by a set of quantum numbers describing their intrinsic properties, such as charge and spin. The stable particles are the electron, proton, photon, and neutrino. The neutron is also stable when bound in an atomic nucleus. The electron is the natural unit of electric charge.

Emission : It is the liberation of electrons or electromagnetic radiation from the surface of a solid or liquid, generally electrons from a metal. The outer electrons of the atoms in a metal (conduction electrons) move in a zag zag manner among the lattice atoms having no net forces on them in the bulk of the material. Electrons near the surface of the material with directions of motion out of the surface may leave the surface and experience a force sending them back to the metal because the metal is left positively charged. The charge on the metal can be considered as an electric image located at the same distance inside the metal as the electron is outside it. The force on the electron varies with distance x from the

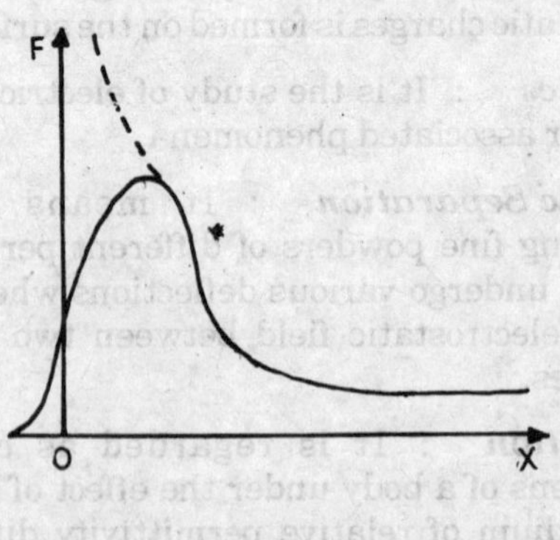

(a) Force on emitted electrons as function of distance from surface.

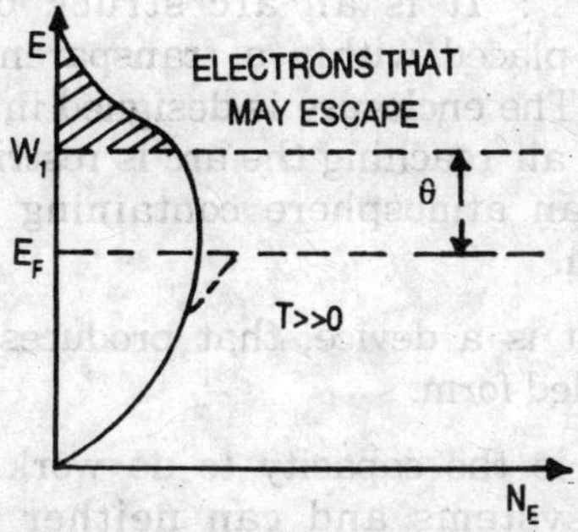

(b) Number of electrons as a function of electron energy.

surface (Fig. a).

Mostly only a few electrons will have velocities, due to their thermal energy, large enough to escape (Fig. b). Emission takes place when sufficient energy is given to the electrons to allow them to escape or the potential barrier is distorted by the presence of an intense electric field.

Emissivity Symbol : When radiation is giving out by the thermal excitation of atoms, molecules, etc., the emissivity is defined as the ratio of the power per unit area radiated from the surface to that radiated by a black body at the same temperature.

Emitter : It is the region of a bipolar junction transistor from which carries flow, through the emitter junction, into the base. The electrode attached to this region is called the emitter electrode.

Emitter-Coupled Logic (ECL) or Current Mode Logic (CML) : It is a family of integrated logic circuit so called because a pair of transistors coupled by their emitters forms a fundamental part of the circuit. It should be noted that the basic ECL gate has simultaneously the function required and its complement.

Emitter Follower : It is an amplifier that consists of a bipolar junction transistor with common-collector connection, the output being taken from the emitter.

Enabling : It means activating a particular circuit or group of circuits from a larger set of circuits in order to effect their operation. An enabling pulse is often used to select the desired circuit.

Enclosed Arc : It is an arc struck between carbon electrodes placed within a transparent or translucent enclosure. The enclosure is designed in such a way that the flow of air reaching the arc is restricted and the arc burns in an atmosphere containing the products of combustion.

Encoder : It is a device, that produces an output in a desired coded form.

Energy : It is the capacity to do work. it exists in all dynamic systems and can neither be created nor destroyed. It can however be transfered from one form to other.

Energy Bands : In a single atom, according to quantum theory, the orbiting electrons can only have certain discrete energies. The atom has a number of associated energy levels and the electrons occupy the lower levels and

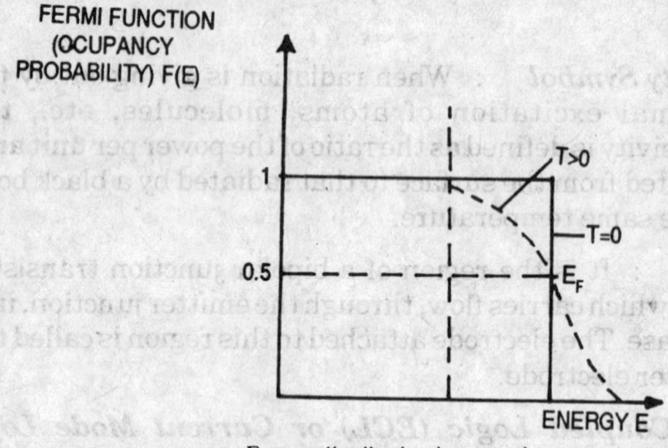

Energy distribution in a metal.

obey the Pauli exclusion principle, i.e., not more than two electrons can occupy each level.

The probability of finding an electron at a particular energy is its Fermi function, F(E). Fig. shows the function for a metal at absolute zero and a temperature above absolute zero.

Energy Levels : They give the possible values of energy of an atom or molecule. Emission or absorption of photons of radiation.

Enhancement Mode : It refers to a means of operating field effect transistors in which increasing the magnitude of the gate bias-increases the current.

ENIAC : Acronyn from Electronic Numerical Integrator and Computer. It is one of the first large-scale digital computers to use electronic techniques. It was built by the University of Pennsylvania and employed about 18000 valves.

Epitaxy or Epitaxial Growth : The process of growing a thin layer of material upon a single-crystal substrate, like silicon, so that the lattice structure of the layer is identical to that of the substrate. The material, which may be the same as the substrate or a different one, is generally deposited from a gaseous mixture. The technique is generally used in semiconductor technology when a layer (the epitaxial layer) of different conductivity to the substrate is required.

Epoxy Resin : It is a type of synthetic resin which is widely used for encapsulating electronic devices due to its properties of high strength and low shrinkage.

Equal Energy Source : It is a source of energy in which the emitted energy is equally distributed throughout the entire frequency range of the source's spectrum.

Equalization : It is a means of reducing distortion in a system by introducing networks that compensate for the particular type of distortion over the required frequency band.

Equivalent Circuit : It is an arrangement of simple circuit elements with the same characteristics as a more complicated circuit or device, under specified conditions. It is noted that it may be used to predict the behaviour of the more complicated system.

Equivalent Network : An electrical network which may replace another network without materially affecting the conditions in other parts of the system, but mostly at one particular frequency only.

Equivalent Resistance : The value of total resistance that in case placed at a point in a circuit, would dissipate the same power as the total of smaller resistances in the circuit.

Equivalent Sine Wave : A sine wave which has the same fundamental frequency and root-mean-square value as a particular periodic wave.

Erase : It refers to remove stored information from a location in storage tubes, memories, magnetic tape, etc.

Erg : The unit of work in the CGS system of units. In S I. units one erg equal 10^{-7} joule.

Etching : it is erosion of selected portions of surface using acids or other corrosive agents in order to produce the desired pattern on the surface. The technique is mostly used in microminiature electronics.

Ettinghausen Effect : It is the effect which gives a very small transverse temperature gradient in a conductor when carrying current in a magnetic field. The temperature gradient is placed in a direction perpendiculor to both the magnetic field and the current.

Eutectic Bond : A thermometallurgical bond which gives contact between somiconductor chips.

Excess Conduction : It is the conduction in a semiconductor due to the presence of electrons which are not required to complete the chemical bonding of the semiconductor and are thus able to conduct charge through the semiconductor. These excess electrons are mostly supplied by a donor impurity.

Excitation : It may be defined as follows:

(i) The addition of energy to an atom or molecule exciting it from the ground state to a higher energy level.

(ii) The application of a signal to the base or control electrode of a transistor or value.

(iii) The application of voltage to an oscillating crystal.

(iv) The application of radio frequency pulses to a tuned circuit.

(v) The application of current to the winding of an electromagnet in order to produce a magnetic flux. The applied current is called the exciting current.

Extrinsic Semiconductor : It means a semiconductor in which impurities or imperfections determine the charge-carrier concentration.

F

Facsimile Transmission : It is a process of transmitting any kind of graphic material to get a pictorial likeness of the object. The system employs *facsimile scanning*: the subject copy is canned, thus providing a successive analysis from which electrical signals are produced. These signals are transmitted to a receiver.

The *facsimile receiver* detects and amplifies the incoming signals then sends them to a light source. The light source is brightness-modulated by the receiver signal in order to record the pricture elements.

Fader : It is type a of device which maintains an electrical signal at a constant level while one signal is being *faded* out and another faded in.

Fading : It is regarded as variations in signal strength at a receiver due to variations in the transmission medium. Destructive interference between two waves travelling by two different paths to the receiver is the most common cause of fading; this is called *interference fading*.

Amplitude fading takes place if all transmitted frequencies are attenuated approximately equally, resulting in a smaller receiver signal. Selective fading takes place when some frequenices are more attenuated than others, resulting in a distorted received signal.

Failure : It is the stoppage of performance of its required function by a device, component, circuit, or any part or subsystem which can be separately tested. The mechanism is the physical, chemical, metallurgical, or other process resulting in a failure.

Failure can occur gradually i.e., unanticipated by prior examination or anticipated. It may also be partial, complete, or intermittent. A failure that is both sudden and complete, is both termed *catasrophic failure*; while one that is both gradual and partial is called *degradation failure*.

Failure Rate : It is defined as the number of failures of an item per unit measure of life. For any particular item the failure rate will be based on the results of life tests and one or more of the following will be applicable.

The observed failure rate is the ratio of the total number of failures in a single population to the sum of the times during which each item has been subjected to stress conditions.

The *assessed failure* rate is known as a limiting value of the confidence interval with a stated probability level based on the same data as the observed failure rate of nominally identical items.

It is to be noted that the *extrapolated failure rate* is an extension of the observed or assessed failure rate by defined extrapolation or interpolation for durations or stress condition which are different from those applying the conditions of the assessed rate.

Fall-Time : It is regarded as the time required for a logic circuit to change its output from a high level (logical I) to a low level (logical 0).

Fan-in : It is defined as maximum number of inputs acceptable by a logic circuit.

Fan-out : The maximum number of inputs to other circuits that the output of a given circuit can drive, within a group of logic circuit.

Farad : Its symbol is F and is the SI unit of capacitance. A capacitor has a capacitance of one farad when a charge of one coulomb increases the potential difference between its plates by one volt. The farad is large for most practical purpose, the sub-multiples microfarad, nanofarad, and picofarad are more convenient.

Faraday Constant : It is a fundamental constant which has the value 9.648670×10^4 coulombs per mole. It is regarded the quantity of electricity that is equivalent to one mole of electrons and that can deposit or liberate one mole of a univalent ion. It that product of electron charge and Avogadro constant.

Faraday's Law of Electromagnetic Induction : It states that if a magnie is moved towards a coil of wire (sole acid) connected in serie in a galvanometer an elective current is produced in the circuts.

"When the magnet is moved towards the solenoid, the galvanometer shows deflection is one direction and when the magnet is moved away fromt the solenid, the galvanometer shows deflection in other direction. When the magnet is stationary there is no deflection in the galvanometer.

The laws are as under:

1. When ever there is a change is magnetic lines of force a magnetic flux, and induced current is produced in the circuit.

2. The induced current or e.m.f. lasts only for the period in which the lines of force or magnetic flux are actually changing.

3. The magnitude of the induced e.m.f. depends upon the rate at which the magnetic lines of force or magnetic flux changes.

Fast-Recovery Diode : It is a diode in which very little carrier storage occurs and that may therefore be used to give an ultra high speed of operation. Fast-recovery diodes can be formed as p-n. junction diodes fabricated from a direct-gap semiconductor. The switching time from forward to reverse bias is of the order of 0.1 nanoseconds or less compared to about one to five nanoseconds in silicon.

It should be noted that the fast recovery diodes may also be formed from Schottky diodes. The carrier storage in a Schottky diode is negligible because they are majority-carrier devices, and a similar speed of operation is achieved.

Fat Zero : It is a significant charge packet usd in charge-coupled devices that corresponds to a zero input sample.

Fault : It is a defect in any circuit or device that interferes with it or prevents it from operating normally.

Fault Current : It is regarded as a current that may flow through a circuit or device as a result of a fault, like a defect in the insulation. The current can take the form of an arc back, short circuit, electrical surge, current to earth, etc.

Feedback : It is defined as the process of returning a fraction of the output energy of an energy-converting device to the input. The circuit which transmits the feedback signal to the input is the *beta circuit*; the circuit containing the active device, which generates the output signal, is learned as the *mu circuit*.

In the case of an active device like as a transistor which introduces a gain A in the absence of feedback, voltage feedback is used if a fraction β of the output voltage is returned to the input (Fig. a). The effective output voltage is given by the equations.

$$v_o = A(v_i + \beta v_o)$$

The overall gain of the combination is then given by the relation

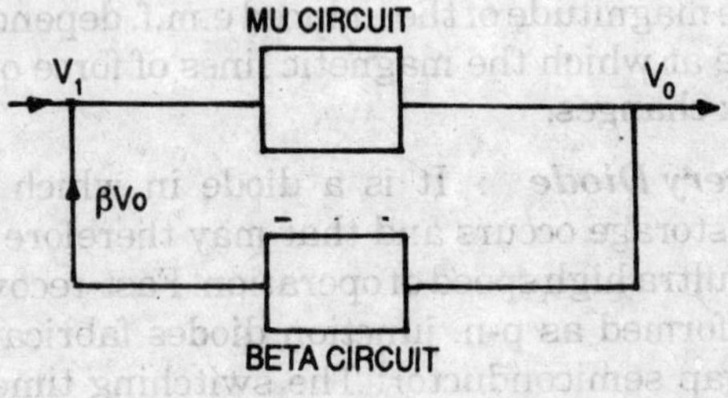

a Voltage feedback circuit

$$vo/vi = A/(1 - \beta A)$$

In case β is negative the feedback voltage opposes the input voltage and the process is termed *negative feedback*. The overall gain of the device is diminished but there is a corresponding reduction in the amount of noise and distortion in the output. In case the term (- βA) is made large compared with unity the overall gain reduces to 1β and is independent of the elements in the mu circuit.

When β is positive the feedback voltage increases the input voltage and the process is termed *positive feedback*. The overall gain of the device is increased and when the factor (β A) becomes equal to or greater than unity the output voltage becomes independent of any input signal and oscillations takes places. The point at which the term (-βA) just becomes unity for any given circuit is termed the singing point.

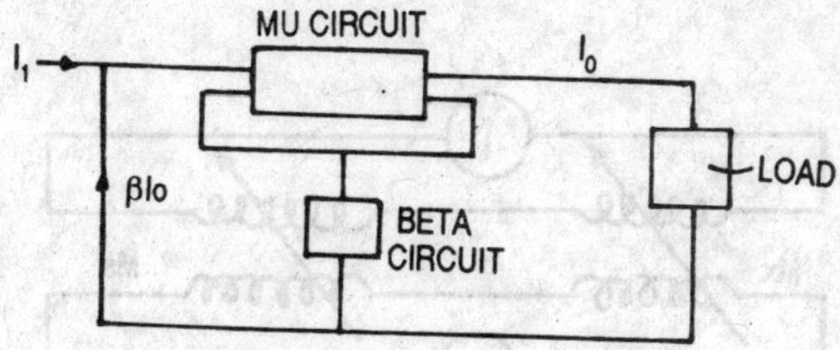

(b) Current feedback circuit

Current feedback is a type of feedback in which a fraction of the current output to the load is feedback to the input (Fig. b). The effective output current is given by the relation

$$i_o=A(i_i+ \beta\, i_o)$$

and the overall gain is given by the equation

$$i_o/i_i=A/(1- \beta\, A)$$

Feedback Control Loop : It is a method used for many different types of control sytem, in which a portion of the ouput derived from a system is feedback to the input circuit in order to control the ouput signal in desired manner.

Feeder : It is the part of a radio system which conveys the radiofrequency energy from the transmitter to the aerial or from the aerial to the receiver.

The term is also used for an electric line that conveys electrical energy from a generating station to a point of a distributing network without being tapped at any intermediate points.

Feedthrough : It is regarded as a contact on a printed circuit board that connects one layer of interconnectians with the next layer, passing through the insulating material that separates them. In an integrated circuit containing multilayer interconnections, feedthroughs may be used to make contact between one layer of interconnections and the next.

Felici Balance : It is a type of alternating-current bridge that is usually used to determine the mutual inductance between the coiling of an inductor. The unknown inductance. M_x is compared to a known standard inducatance

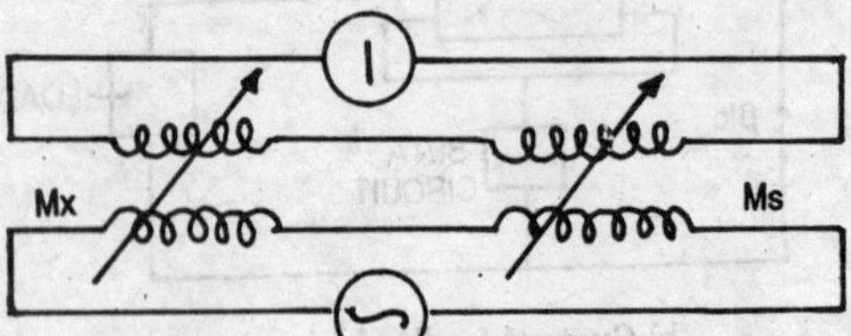

Felici Balance

M_s (see diagram). At balance, when there is no response from the indicating instrument, I we can write

$$M_x = -M_s$$

Fermi Dirac Statistics : It is regarded as a system of quantum statistics which is used to describe the behaviour of solids in terms of a free electron model.

The probability of an enrgy level of energy E being occupied by an electron is given by the *Fermi-Dirac distribution* function, $f(E)$ as follows

$$f(E) = 1/\left[e^{(E-\mu\, K^T)} + 1\right]$$

where μ is a constant and a function of temperature, k is the Boltzmann constant, and T the thermodynamic temperature.

The free electron model is utilized to explain a number of important physical properties of metals.

Fermi Level : Symbol: E_F. This is defined as the maximum electronic energy level which is occupied by an electron in a solid at a temperature of absolute zero. At higher temperature some electrons are excited into higher energy states. It should be noted that the Fermi level then corresponds to the value of energy at which the Fermi-Dirac distribution function has a value $\frac{1}{2}$.

Ferrimagnetism : It is effect observed in solids, notably garnets and ferrites, in which the magnetic properties change at a certain critical temperature knwon as the *Neel temperature.*

At temperatures below the Neel temperature ferrimagnetic materials behave in a similar manner to ferromagnetic materials; they show spontaneous magnetization within a domain structure and magnetic hysteresis, but the spontaneous magnetization is less than that of ferromagnetic materials and does not correspond to full parallel alignment of the individual magnetic moments.

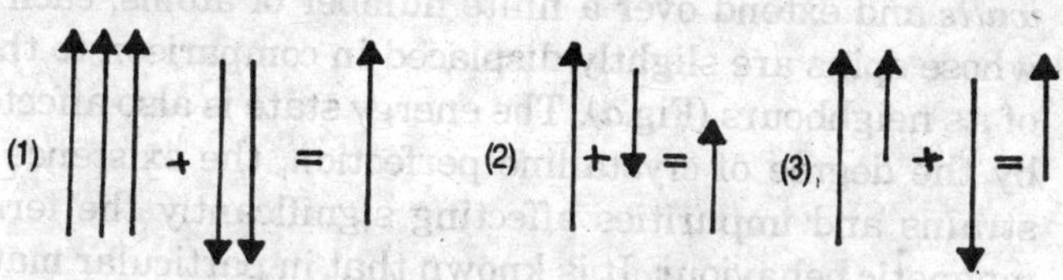

Possible arrangement of magnetic moments on two sublattices

The antiparallel spins are unequal, a net magnetization. Some possible arrangements are shows in the diagram: (1) unequal numbers of identical moments; (2) unequal moments; (3) two identical moments on each sublattice plus one unequal moment on one. It should be noted that the net small spontaneous magnetization results in the material to behave as a weak ferromagnetic material.

Ferroelectric Crystals : These are crystals which show electrical properties analogous to certain magnetic properties, like feromagnetism. In an alternating electric field very large values of the piezoelectric and dielectric constants are developed, generally in one particular direction within a certain temperature range. These crystals are mostly useful as detectors of vibrations.

Ferromagnetism : This is regarded as a phonemenon observed in certain solids in which the magnetic properties change abruptly at a cetain characteristic tempreature which is temed as the *Curie point*. Below the Curie poin the solid exhibits ferromagnetic properties. Above this temperature the thermal energy of the atom is enough to produce magnetic properties typical of paramagnetism.

The magnetic moments of the atoms is due to the spin of electrons in an unfilled inner shell. In any stable material in the absence of an external magnetic field, the, detailed arrangements of the magnetic moments is due to the interaction between the various forces operating within the sample. It always produces the minimum energy possible. In ferromagnetic materials this minimum energy state takes place when the electron spins of the atoms within a domain are arranged in parallel.

The domain size is determined by a compromise between competing forces. In boundaries are called *Bloch walls* and extend over a finite number of atoms, each of whose spins are slightly displaced in comparison to that of its neighbours (Fig.*a)*. The energy state is also affected by the degree of crystalline perfection, the existence of strains and impurities affecting significantly the ferromagnetic behaviour. It is known that in particular material the energy state of the domains is least when the Bloch wall intersects as many dislocations as possible. A typical energy curve as a function of wall position is depicted in Fig.*b*. A virgin specimen would have a well located at the minimum energy position, marked A.

(a) Representation of spin direction in a Bloch wall with 180 rotation

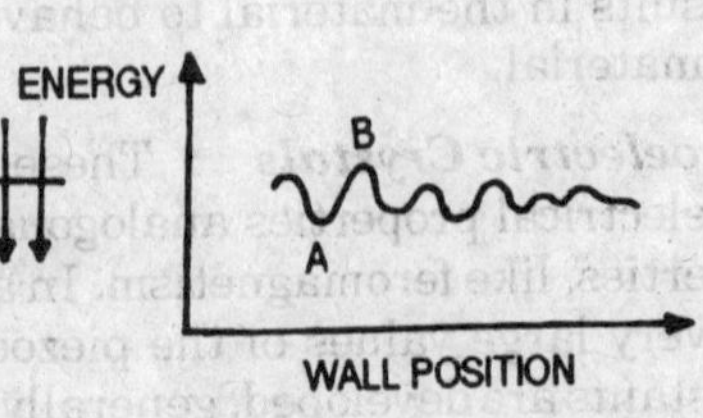

(b) Energy as a function of bloch wall position

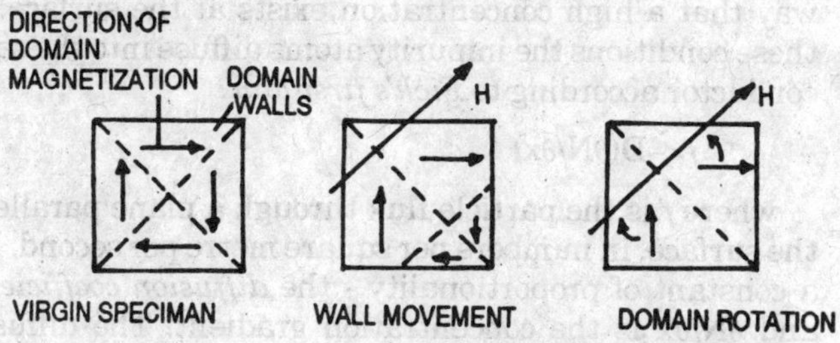

(c) Representation of magnetization of a ferromagnetic sample

At low value of magnetic field the net effect is to displace the Bloch walls over a few atoms, away from the minimum energy state; which means those domins with spins parallel or nearly parallel to the field grow at the expense of the others (Fig.*c*). In case the field is removed the walls tend to move back to the minimum energy state and for small values of applied field the magnetization changes are small and reversible.

Fetron : It is called a junction field-effect transistor mounted in a suitable package so that it can directly replace a valve in a circuit without any change in the circuit.

Fibre-Optics System : This is a system that uses one or more glass or perspex fibres as a *light guide* or for transmitting optical images. The polished surfaces of fibres are coated with a material of suitable refractive index. Light entering one end within a certain solid angle undergoes total internal reflection at the surface and is transmitted through the fibre. When the critical angle of the material is not exceeded the fibres may be curved.

Fick's Law : It states that in a concentration gradient of mobile impurity atoms exists in a semiconductor there will be a flow of such atoms from the region of high concentration to the region of low concerntration. This effect is used in producing a desired impurity profile in a particular specimen of semiconductor.

At normal temperatures the impurity atoms are immobile till heated to a high temperature. A concentration gradient is obtained by heating the semiconductor water in a gaseous atmosphere of the impurity atoms in such a way that a high concentration exists at the surface. By these conditions the impurity atoms diffuse into the semiconductor according to *Fick's first law* :

$$f=-D(\partial N/\partial x)$$

where f is the particle flux through a plane parallel to the surface, in numbers per square metre per second. D is a constant of proportionality - the *diffusion coefficient* - and $\partial N/\partial x$ is the concentration gradient. The diffusion coefficient, D, is given by the relation

$$D=\mu k\ T/e$$

where k is the Boltzmann constant, T the thermodynamic temperature, e the electron charge, and m the mobility of the impurity atoms. As N is also a function of time during the diffusion process, *Fick's second law* can be derived from the first law and is given by

$$\partial N / \partial t=D(\partial^2 N / \partial x^2).$$

Field : The field is the region in which the physical agency exerts its influence. Typical examples are electric and magnetic fields resulting from the presence of charge or magnetic dipoles.

Field Control : A method adjusting the voltage of a generator or the speed of a motor by adjusting current in the field coil.

Field-Effect Transistor (FET) : It is a system in which current flows through a narrow conducting *channel* between two electrodes and is modulated by an electrical field applied at a third electrode.

There are two main kinds of FET the *junction FET* (JFET, JUGFET), in which the conducting channel forms part of the structure of the device, and the *insulated gate FET* (IGFET), in which the channel is obtained in use by the action of the gate voltage. It is illustrated in Fig.*a*.

The junction FET consists of a water of semiconductor material flanked by two highly doped layers of the opposite conductivity type (n^+ or p^+), forming the source, drain, and gate regions (Fig.*b*).

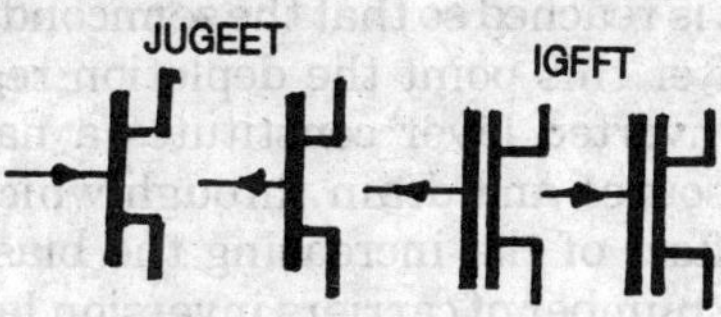

n-channel p-channel n-channel p-channel
a-symbols for FET types

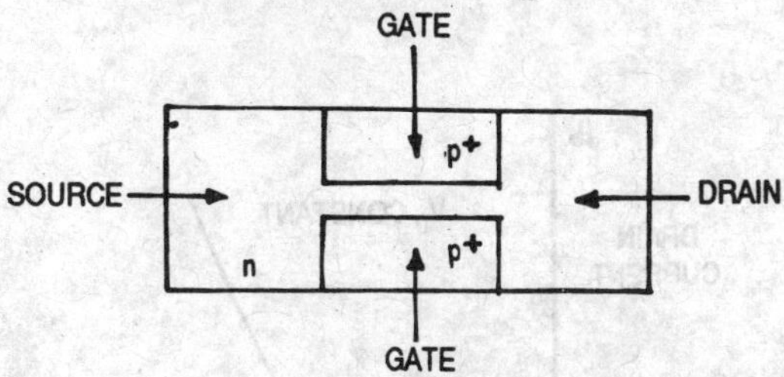

b-JUGFET

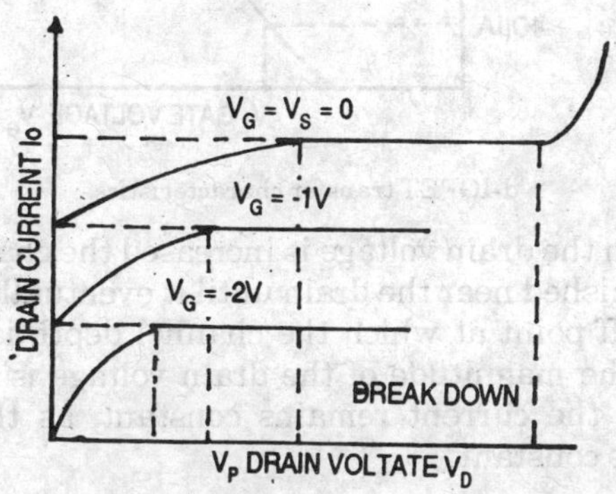

c-Characteristic curves of a JUGFET

Using a negative voltage to the gate will also increase the size of the depletion layers and thus the pinch-off condition will reached at a lower drain voltage. A family of characteristics will thus be generated for different values of gate bias (Fig.*c*). It means that the gate voltage is therefore used to modulate the channel conductivity.

In case a positive voltage is applied to the gate electrode of an *n*-channel device, a depletion layer is obtained in the substrate below the insulating layer. With the increase of bias the depletion layer spreads into the semiconductor until a point is reached so that the semiconductor becomes inverted. After this point the depletion region ceases to grow. The inverted layer constitutes a narrow channel connecting source and drain through which current can flow. The effect of the increasing the bias further is to increase the number of carriers inversion layer and hence the conductivity of the channel. The gate voltage at which strong inversion takes place is the *threshold voltage*, V_T (Fig.*d*).

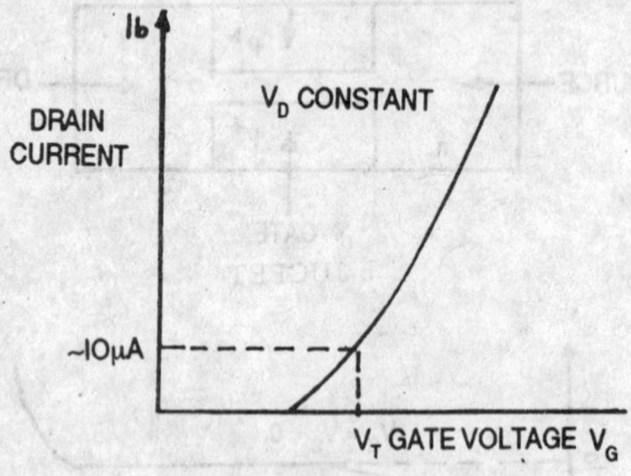

d-IGFET transfer characteristics

When the drain voltage is increased the channel depth is diminished near the drain until it eventually reaches a pinch-off point at which the channel depth is just zero. When the magnitude of the drain voltage is further increased the current remains constant, as the channel remains constant.

Field-effect transistors can be broadly described as square-law devices, *i.e.*, the output current I_{DS} varies with the square of the input voltage V_{GS}. The basic device equations are given by : Junction FET (beyond pinch-off):

$$I_{DS}=I_{Dss}[VI\text{-}_{GS}/V_P]^2$$

where I_{DSS} is the saturation drain current at V_G=O and V_P is the pinch-off voltage. The pinch-off condition is represented by

$$V_{DS}=V_{GS}-V_P$$

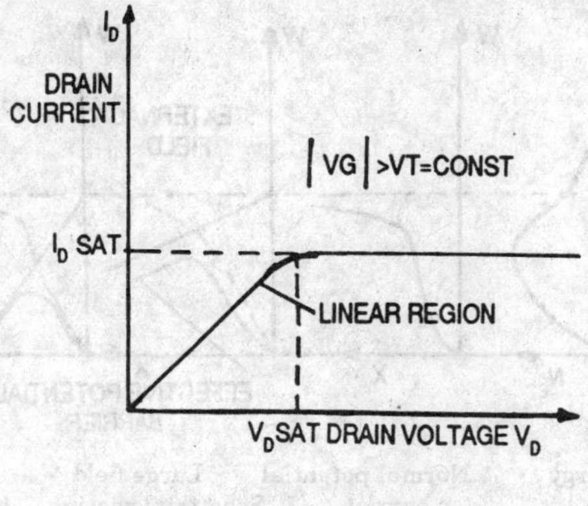

e-Drain-current characteristic

Insulated-gate FET (below pinch-off) is given by the equation

$$I_{DS}=k[2(V_{GS}-V_T)V_{DS}-V_{DS}^2]$$

where k is the gain factor and V_T is the threshold voltage. IGFET (beyond pinch-off) :

$$I_{DS}=k(V_{GS}-V_T)^2$$

The pinch-off condition is given by the equation

$$V_{DS}=V_{GS}-V_T$$

The mutual conductance, $\partial IDS/\partial VG$, of both types of FET is proportional to the square root of the drain current, *i.e.,*

$$g_M=\partial I_{DS}/\partial V_G \propto \sqrt{I_{DS}}$$

Field Emission : This is regarded as an emission in which the presence of a large external accelerating electric field decreases the potential barrier at the surface of the emitter and permits electrons to escape from the surface. The potential barrier is diagram as the work done, W, for an electron to escape.

The distortion of the potential barrier at very large values of the field gives an effective narrowing of the barrier and permits the tunnel effect to operate : electrons with energies around the Fermi level may also be emitted.

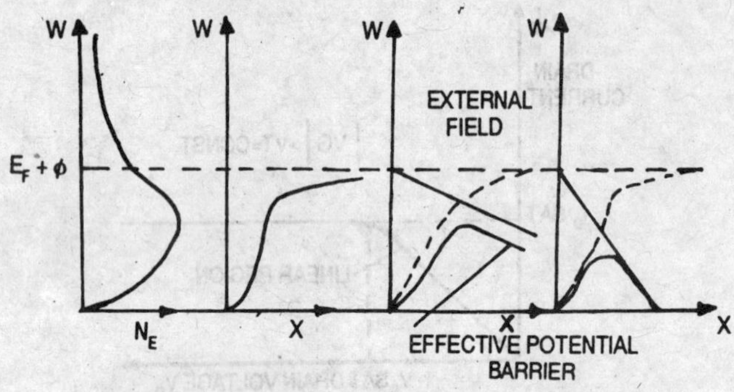

Electron energy distribution | Normal potential barrier | Large field : Schottky lowering | Intense field : tunnel effect

The current density j has been shown to vary with the electric field E and is given by the equation

$$j = CE^2 exp(-D/E)$$

where C and D are approximately constant. The very intense fields required are of the order of 10^{10} volts per metre and are usually only obtained at sharp points on the emitter surface.

Field-Emission Microscope : The microscope is an instrument used for studying the surface of a solid, usually a metal, by causing it to undergo field emission.

Field-Enhanced Emission : This is regarded as the increase in photoemission and secondary emission in the presence of a strong electric field which is applied at the surface of the emitter.

Field Ionization : This is defined as a process in which electrons are emitted from an atom in the presence of an intense.

Field-Ion Microscope : An instrument used for studying the surface of a solid, generally a metal, by subjecting it to field ionization. The form is identical to the field-emission microscope but the voltage is applied in the opposite direction.

Field Magnet : It is the magnet used to provide the magnetic field in magnetic circuits and machines. An electromagnet is generally used but small electrical machines use a permanent magnet also.

Filament : It is defined as a thread-like body of metal or carbon, mostly the conductor of an incandescent lamp, the cathode of a thermionic value, or the electrode applied to heat an indirectly heated cathode.

Film : It is regarded as an electron coating with a minimal thickness dimension. Thin films have thickness in the range 10 nanometres to one micrometre while the thick films range from 10 to 100 micrometres.

Film Resistor : It is a type of resistor that has a thin layer of resistive material deposited on an insulating core. It is to be noted that for low-power applications film resistors are more stable than composition resistors and except for very high precision requirements are smaller and less expensive than accurate wire-wound resistors.

Resistive materials used are crystalline carbon, boron-carbon, and various metallic oxides or precious metals. Film resistors mostly have a continuous uniform film applied in a particular pattern to the core, the film thickness determining the resistance.

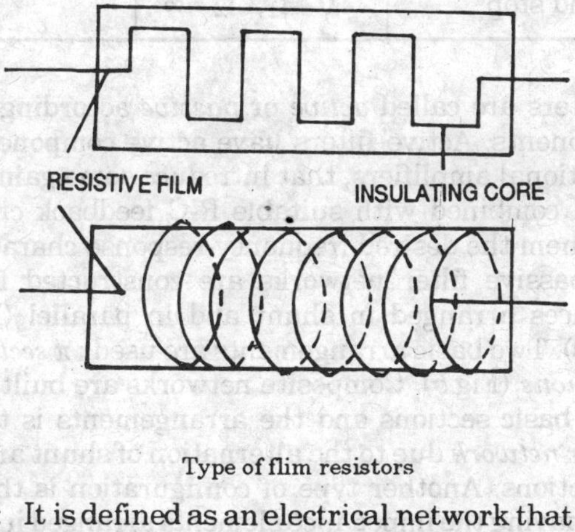

Type of flim resistors

Filter : It is defined as an electrical network that will transmit signals with frequencies within certain designated ranges (*pass bands*) and suppress signals cf other frequencies (*attenuation bands*). Filters are classified as *low pass, high pass, band pass,* and *band-stop filters;* the four main classifications and their corresponding frequency limits are shown in the table below.

It should be noted that an ideal filter would transmit the pass band without attenuation and completely suppress the attenuation band, with a sharp cut-off profile. Practical filters however do attenuate the pass band, due to absorption, reflection, or radiation, which gives a loss of signal power; neither do they completely suppress the attenuation bands. A typical curve of output voltage with frequency is depicted in Fig.*a* for a simple low-pass filter: V_P is the peak voltage and V_m is the maximum voltage of an ideal filter.

Type of filter	Pass band (s)	Attenution band(s)
low pass	0—fσ	fc—∞
high pass	fc—∞	0—fc
band pass	f_1—f_2	0—f_1, f_2—∞
band stop	0—f_1, f_2—∞	f_1—f_2

Filters are called *active* or *passive* according to their components. Active filters have active components, like operational amplifiers, that introduce some gain into the signal combined with suitable R-C feedback circuits to give them the desired frequency-response characteristic. The passive filter networks are constructed from impedances arranged in shunt and in parallel (L-C networks). Two basic arrangements are used : π *sections* and T-*sections* (Fig.*b*). Composite networks are built up from these basic sections and the arrangements is termed a *ladder network* due to the alternation of shunt and parallel sections. Another type of configuration is the *lattice filter* having the impedance elements arranged in a bridge network (Fig.*c*).

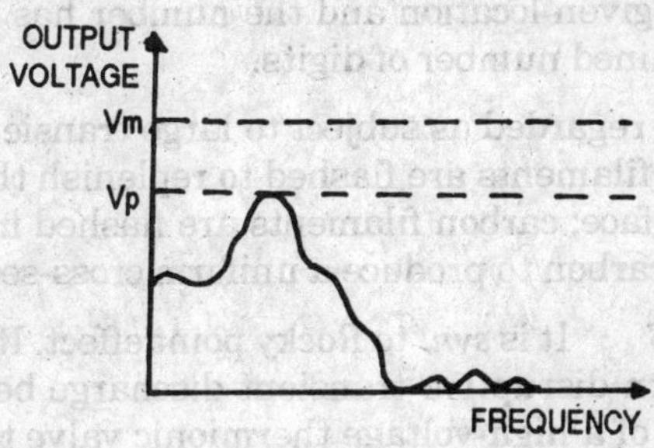

a-Low-pass filter output

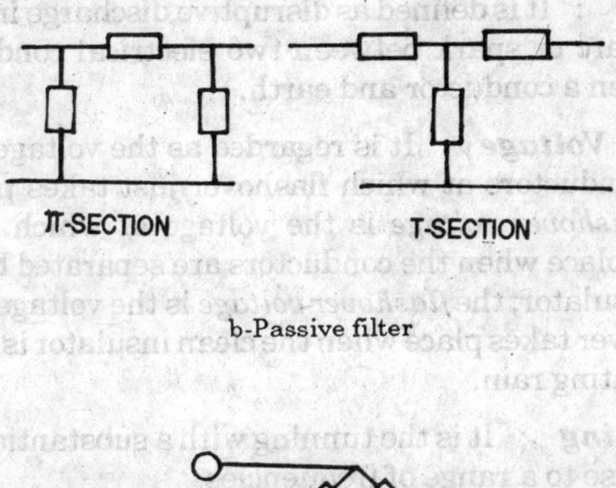

b-Passive filter

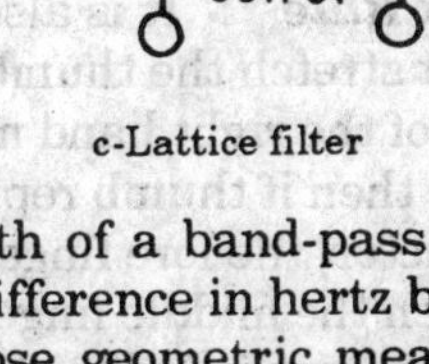

c-Lattice filter

The bandwidth of a band-pass or band-stop filter is defined as the difference in hertz between two particular frequencies whose geometric means equals that of the geometric mid-frequency of the pass or attenuation band.

Fixed-Point Representation : This is a system of representing numbers in which the decimal or binary point is fixed in a given location and the number has a constant predetermined number of digits.

Flash : It is regarded as subject to large transient voltage. Thoriated filaments are flashed to replenish the thorium on the surface; carbon filaments are flashed in an atmosphere of carbon to produce a uniform cross-section.

Flash arc U.S : It is *syn.* to Rocky point effect. It is defined as a suddrn disruptive transient discharge between the electrodes of a high-voltage thermionic valve that generally but not invariably results in the destruction of the valve.

Flashback Voltage : It is defined as the peak inverse voltage across a gas-discharge tube needed to produce ionization.

Flashover : It is defined as disruptive discharge in the form of an arc or spark between two electrical conductors or between a conductor and earth.

Flashover Voltage : It is regarded as the voltage between two conductors at which flashover just takes place. The *dry flashover voltage* is the voltage at which flashover takes place when the conductors are separated by a clean dry insulator; the *flashover voltage* is the voltage at which flashover takes place when the clean insulator is wet, thus simulating rain.

Flat Tunning : It is the tunning with a substantially equal response to a range of frequencies.

Fleming's Rules : It is used for finding the directional relation between current, motion and magneted field in a dynamo or in electric motor.

Fleming Right Hand Rule : It is also known a generator rule. It states that stretch the thumb, the fore finger and the middle finger of the right hand mutually perpendicular to one another then if thumb represents the direction of motion of conductor, the fore finger the direction of the magnetic field, then the middle finger points in the direction to which the current is induced in the circuit.

Flexible Resistor : This is called a type of wire-wound resistor in which the wire is wound round a flexible insulating core.

Flicker : It is defined as the visual perception of brightness fluctuations of less than 25 to 30 per second, in other words less than the persistence of vision. The threshould for perceiving flicker is dependent on the brightness of the observed light and the angle it subtends to the optic axis. In television flicker prevents complete continuity of the images. In colour television flicker can result from either undesirest variations in the luminance signal-*luminance flicker-or in the chrominance signal-colour flicker*.

Flip Chip : It is defined as a semiconductor chip with thickened and extended bonding pads enabling it to be flipped over and mounted upside down on a suitable substrate, such as a thin-film or thick-film circuit.

Flip-Coil : It is a method of measuring the value of magnetic flux density, B, at a point in air. The flip-coil consists of a number of turns, N, wound on a small former of area A.

Flip-coils can be constructed with different values of NA (turns × area) : the appropriate coil can be chosen for the field to be measured, in such a way to produce a suitable instrument deflection.

Flip-Flop : It is *syn.* to half-shift register. It is defined as a bistable multivibrator circuit which generally has two inputs corresponding to the two stable states. It is because application of a suitable input pulse results in the device to 'flip' into the corresponding state and remain in that state until a pulse on the other input causes it to 'flop' into the other state.

It is to be noted that flip-flops are widely used in computers as counting and storage elements and several types have been developed.

Floating : It is described as a circuit or device which is not connected to any source of potential.

Floating-Carrier Modulation : This is defined as a amplitude modulation where in the amplitude of the carrier wave does not remain constant but is automatically varied in a manner depending on the amplitude of the modulating wave, which is averaged over a short time period. The modulation factor thus remains substantially constant.

Floating-Point Representation : It is regarded as method of representing numbers with the help of a predetermined number of significant digits-the *mantissa*-together with a decimal or binary multiplier- the *exponent* : the number x can be expressed as

$$x=y \times n^z$$

where y is the mantissa, n is either 10 or 2, and z is the exponent in integer form.

Flourescent lamp : It is defined as a lamp in which light is generated by fluorescence. A common form of flourescent lamp has a gas discharge tube containing a low-pressure gas, such as mercury, with the inner surface of the tube coated with a phosphor. When a current passes through the tube the ultraviolet radiation obtained strikes the phosphor, as a result of which then emits visible radiation. The usual sodium vapour or mercury vapour street lamp, does not have a flourescent coating.

Fluorescent Screen : A screen used in various electronic devices, to convert the electron beam into a visible image. These screens are made of an array of many small-diameter (two to three micrometre) phosphor crystals that emit light when bombarbed by high-energy radiation, such as *X*-rays, or by electrons.

Flux : This is a measure of the strength of a field of force through a specified area.

Flux Density : It is defined as number of flux lines per unit area in the magnet field.

Fluxmeter : This is an instrument used to measures changes in magnetic flux. The most common type is the *Grassot fluxmeter,* which consists of a moving-coil galvanometer designed in such a way that the restoring couple on the moving coil is negligibly small and electromagnetic damping is large. It should be kept in mind that the galvanometer is used in conjunction with an exploring coil of known area. A change in the magnetic flux cutting the exploring coil results in an induced current in the galva-

galvanometer is used in conjunction with an explcring coil of known area. A change in the magnetic flux cutting the exploring coil results in an induced current in the galvanometer coil and hence the latter is deflected. The angle of deflection is directly proportional to the change in magnetic flux through the exploring coil.

Flying-Spot Scanner : A device which gives a video signal from an object, like a film, by scanning the object with a spot of light, which is then focused on a photocell to produce corresponding electrical signals. The moving (or 'flaying') spot of light is normally obtained on the screen of a high-intensity cathode-ray tube used as a light source.

Flywheel Effect : The continuation of oscillations in an oscillator during the intervals between exciting pulses is termed as fly-wheel effect.

FM : It is an abbrevation for frequency model atom.

FM Receiver : It is a process of making a beam of radiation or particles converge. In ann electron-beam device, like a cathode-ray tube, three methods of focusing the beam are used as given below

In *electrostatic focusing* two or more electrodes as different potentials are used to focus the electron beam.

In *electromagnetic focusing* the action of magnetic field is used to make the electron beam converge.

In *gas focusing* a small amount of an noble gas like argon or helium, is taken into the tube.

Foil Capacitor : It is termed as a capacitor in which the electrodes are metal foil. The term is generally used for paper capacitor by some polystyrene or polyester film capacitors use foil electrodes and one form of tantalum electrolytic capacitor employs a tantalum foil as one of the electrodes.

Force : It is defined as that which produces or tends to produce motion or change of motion. In mechanical engineering it is push or pull applied on any body.

Forced Oscillation. : It is defined as the oscillations produced in a circuit which is acted upon by an external driving force, oscillations in a reasonant circuit coupled to a fixed-frequency oscillator. The resulting oscillations have two components: a transient component, whose frequency is determined by the natural frequency of the circuit and decays rapidly, and a steady component, whose frequency equals to that of the external driving force.

Force Factor : It is regarded as the ratio of the force required to block the movement of an electromechanical transducer to the corresponding current in the electrical system.

Form Factor : It is the ratio of the root-mean-square value of an alternating quantity (such as current or voltage) to the half-period mean value, for a half-period beginning at a zero point. A simple sine wave contains form factor equal to $\pi/2\sqrt{2}$ (*i.e.*,1.111).

Forward Direction : It is regarded as that direction in which an electrical or electronic device has the smaller resistance. A voltage applied in a forward direction is known as the *forward bias;* it produces the larger current, known as the *forward current.*

Fourier Analysis : This is a method of analysing complex waveshapes or signals into a series of simple harmonic functions, having frequencies which are integral multiples of the fundamental frequency.

The Fourier series may also be written as a series of sines and cosines or as a series involving the phase angles of the various components.

Fourier analysis is widely used in the electrical communication field.

Four-Terminal Resistor : It is a defined as a standard resistor with four terminals. Two are used to connect the resistor to the current source and the other two to connect it to a measuring instrument. This type of arrangement ensures that the potential drop across the resistor is not affected by contact resistances at the terminals.

Four-Wire Circuit : It is defined as a circuit that has two pairs of conductors and that forms a simultaneous two-way communication channel between two points of a telecommunication system. One pair of conductors forms the 'go' channel and the other the 'return' channel.

It is also one cycle of a number of pulses which regular recur in a pulse train used in pulse train communication.

Franklin Symbol Fr. : This is a unit of charge in the CGS electrostatic system of units. One franklin equals 3.336 × 10^{-10} coulomb.

Free Electron : It is an electron which is not bound to a specific atom or molecule and is thus free to move when influenced by an applied electric field.

Free Field : It is defined as a field in which any boundary effects are negligible in the region of interest.

Free-Field Calibration : Calibration of a microphone in which the open-circuit voltage produced by a certain value of sound pressure in the free wave is determined. The presence of the microphone upsets the free field which existed before the introduction of the microphone, resulting in difficulties in determining the pressure value.

Free Oscillations : Oscillations taking place in a circuit under the influence of internal forces, like a capacitor discharging through a resistance and inductance, or of a constant external force, like a direct voltage. Both these conditions are similar to a mechanical vibrating system being displaced from the neutral point.

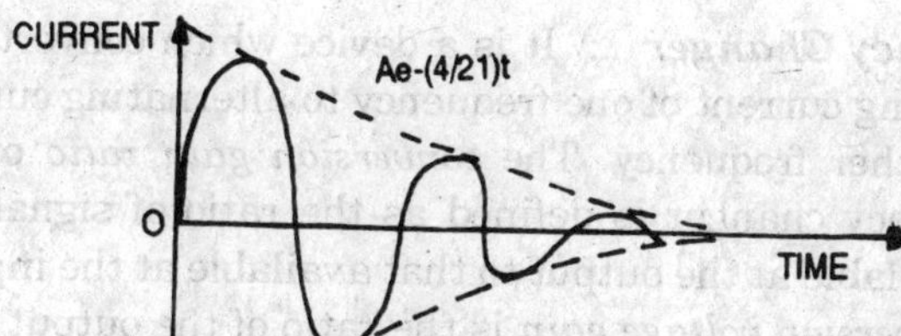

Decay of free oscillations

The oscillation decay slowly, depending on the amount of damping in the circuit, with a frequency, *f*, termed the natural frequency, which is approximately equal to $(LC)^{-1/2}$, in case resistence R in the circuit is small.

Free-Space : It is defined as the region used as an absolute standard and characterized by an absence of gravitational and electromagnetic fields. It was previously called as a vacuum. The velocity of light in free space is constant and is the maximum possible value.

Frequency : Its symbol, is v or f; and unit : hertz. The number of complete oscillations or cycles of a periodic quantity taking place in unit time. The frequency is related to the angular frequency ω by the relation $\omega = \pi v$. The frequency of a periodic quantity, like an alternating current, is given by the number of times the quantity passes through its zero value in the same sense in unit time.

The frequency of electromagnetic radiation is related to the wavelength, λ, by the equation $v = c/\lambda$, where c is the velocity of light.

Frequency Band : It is regarded as a particular range of frequencies that forms part of a larger continuous series of frequencies.

Frequency Bridge : It is defined as an alternating-current bridge whose balance point is dependent on the frequency at which the measurement is carried out.

Frequency Changer : It is a device which converts alternating current of one frequency to alternating current of another frequency. The *conversion gain ratio* of a frequency changer is defined as the ratio of signal power available at the output to that available at the input; the *conversion voltage gain* is the ratio of the output voltage to the input voltage.

Frequency Discriminator : It is a discriminator which selects input signals of constant amplitude and produces an output voltage proportional to the amount that the input frequency differs from a fixed frequency. Frequency discriminators are generally used in automatic frequency control systems.

Frequency Divider : It is a device that produces an output signal whose frequency is an exact integral submultiple of the input frequency.

Frequency Doubler : It is a frequency multiplier which produces an output signal with a frequency twice that of the input signal.

Frequency Meter : The term is used for a an instrument that is used to measure the frequency of an alternating current. The frequency of an electromagnetic wave is commonly measured using a cavity reasonator.

Frequency Modulation (f.m. or FM) : This is a type of modulation having the frequency of the carrier wave which is varied above and below its unmodulated value by an amount proportional to the amplitude of the signal wave and at the frequency of the modulating signal, the amplitude of the carrier wave remaining constant.

Frequency modulation has several advantages over amplitude modulation, the most important being the improved signal to-noise ratio.

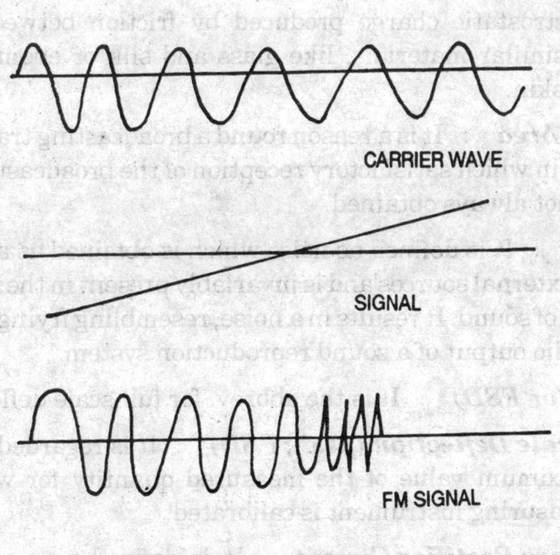

Frequency Modulation

Frequency Multiplier It is regarded as a device which produces an output signal whose frequency is an exact integral multiple of the frequency of the input. Some cases of frequency multipliers are frequency doublers and triplers.

Frequency Range : This is the range of frequencies at which a circuit or device operates normally.

Frequency-Response Characteristic : It is the variation with frequency of the transmission loss or gain of any apparatus, circuit, or device. A series of tests at different frequency designed to determine the frequency-response characteristic is known as a *frequency run.*

Frequency Selectivity : It is the ability of any circuit or device to differentiate between signals at different frequencies or between a signal at a particular frequency and interference at different frequencies.

Frequency Standard, Primary : The term is used for a very stable and precise oscillator that may be calibrated against national standard frequencies and used as a laboratory standard.

Frictional Electricity : The phenomena associated with electrostatic charge produced by friction between two dissimilar materials, like glass and silk or ebonite and catskin.

Fringe Area : It is a reason round a broadcasting transmitter in which satisfactory reception of the broadcast signal is not always obtained.

Frying : It is defined as noise which is obtained as a result of external sources and is invariably present in the recording of sound. It results in a noise, resembling frying, in the audio output of a sound reproduction system.

F.S.D. (or FSD) : It is the abbrev. for full-scale deflection.

Full-Scale Deflection (f.s.d.; FSD) : It is regarded as the maximum value of the measured quantity for which a measuring instrument is calibrated.

Full-wave Rectifier Circuit : It is defined as a rectifier circuit which rectifies both the positive and negative half-

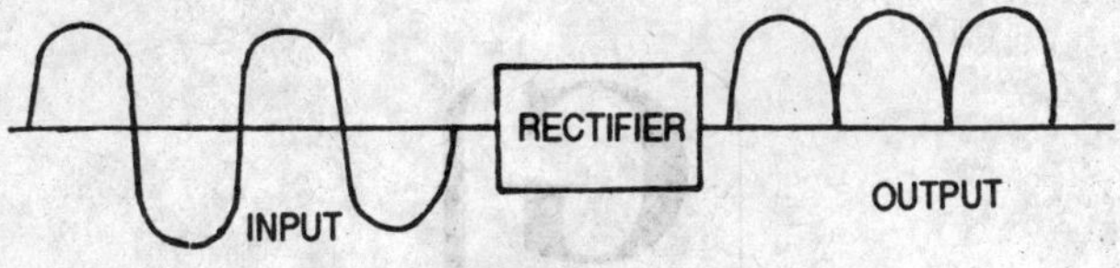

Full-wave Rectifier Circuit

cycle of the single phase a.c. input and delivers them as unidirectional current to the load.

Function Generator : It is a signal generator which gives various specific waveforms for test purposes over a wide range of frequencies.

It is also used as a unit in an analog computer which gives output signal corresponding to the value of a specified function of the independent variable input.

Fundamental Frequency : It is the frequency of a sinusoidal component of a periodic quantity which has the same period as the periodic quantity.

Fuse : It is a short length of easily fusily fusible wire used to protect electric circuits or devices by melting ('blowing') at a specific current and thereby breaking the circuit. The *fuse current rating* is the maximum value of current which the fuse will conduct without melting. The frequency and voltage at which a fuse is designed to operate are specified by the *fuse frequency rating* and *fuse* voltage rating. The *fuse characteristic* is the relation between the current through the fuse and the time taken for the fuse to operate.

G

Gain : Its unit is decibel. It is defined as a measure of the ability an electronic circuit, device, or apparatus to increase the magnitude of a given electrical input parameter. In case of power amplifier the gain is the ratio of the power output by the amplifier to the power input to it. Thus for a voltage amplifier which supplies relatively little power to the load, the gain is the ratio of the voltage developed across a specified load impedance to the input voltage.

Gain Control : It is a circuit or device that varies the amplitude of the output signal of an amplifier.

Gallium Arsenide : A direct-gap 3-5 semiconductor consisting of a relatively large band gap and high carrier mobility. The relatively high carrier mobility permits the semiconductor to be used for high speed applications and due to the large energy gap it has a high resistivity that allows easier isolation between different areas of the crystal.

Gallium arsenide cannot be used since diffusion of impurities into the material is extremely difficult.

Galvanomagnetic Effect : It is a phenomena arising when a current is passed through a conductor or semiconductor in the presence of magnetic field.

Galvanometer : It is an instrument that detects and measures small currents. Commonly used type is the *moving-coil galvanometer,* having a small coil suspended between the poles of a permanent magnet. When the current to be measured is passed through the coil a couple acts on the coil forcing it to rotate in the magnetic field. Before passing the current the suspension is adjusted in such a way that in most types, the coil sets along the direction of the magnetic fields; some galvanometers have the coil set perpendicular to the field. In case a current, i is passed through the coil, the couple and hence the angular deflection, θ, is proportional to $iB \cos \theta$, where **B** is the magnetic flux density. In practice the poles of the magnet are shaped and have a soft iron core between them thus the magnetic field is radial and always parallel to the plane of the coil whatever the deflection; thus cos θ=and $\propto$i.

The deflection may be observed by means of a light non-magnetic pointer which moves over a scale of generally by using a small mirror attached to the suspension. A beam of light is reflected by the mirror and the spot of light moves across a linear scale. This arrangement is sometimes called a mirror galvanometer. The moving-coil galvanometer is virtually independent of the earth's magnetic field, which is very weak compared to the field of the magnet.

The *moving magnet galvanometer* is now no longer used as its disadvantage is that it must be very carefully aligned along the magnetic meridian and is easily disturbed by external magnetic fields. It consists of a needle shaped small magnet, pivoted at the centre of a fixed coil, which is bigger in comparison to the magnet. The magnet is under the influence of a uniform magnetic field, usually that of the earth, and is set up in such a manner that in the zero current situation it is in the plane of the coil (Fig.*b*). A long light nonmagnetic pointer made of aluminium, is attached perpendicularly to the magnet, in such a manner that it swings over a scale. When a current, t is passed through the coil the magnet is forced to rotate and the angular deflection, b, is measured. The earth's magnetic field provides the necessary retarding force. At equilibrium $i \propto \sin \beta$.

Galvanometer Constant : it is defined as the multiplying factor which must be applied to the scale reading of a galvanometer as to give the value of the current in amps.

Gamma Rays (γ-rays) : It is regarded as very high frequency electromagnetic radiation which is emitted spontaneously by certain radioactive elements in the course of a nuclear transition or can be produced in nuclear reactions, as in the annihilation of an elementary particle and its antiparticle.

Gamma rays are different from alpha and beta particles by their greater depth of penetration and their nondeflection in electric and magnetic fields. The depth of penetration is controlled by their energy, which depends on the wavelength.

Ganged Circuits : Two or more circuits having variable elements which are mechanically coupled in such a manner that they can be operated simultaneously by a single control are called ganged circuit.

Ganging Oscillator : It is oscillator that has a constant output over a wide frequency range and whose frequency may be rapidly adjusted. It is helpful for testing the accuracy of adjustment of ganged circuit.

Gap : It is defined as an air gap in a magnetic circuit that increases the inductance and saturation point.

It is also regarded as the space between the electrodes in any electron tube. A spark gap is a special arrangement of the electrodes such that a spark takes place if the voltage exceeds a predetermined value. A spark gap is often used to divert high-voltage surges and thereby protect a device.

Gas Breakdown : A breakdown which takes place in a gas-filled tube when the voltage reaches a given value. Ions in the gas may be accelerated by the field and reach high kinetic energies.

Gas Electrode : It is defined as an electrode which absorbs a gas so that when in contact with an electrolyte the gas effectively acts as the electrode. A *gas cell* is one which has a gas electrode.

Gas-Field Radiation-Detection Tubes : The gas-filled tubes that operates by virtue of the ionization produced in them when charged particles pass through them.

Gas-Filled Tube : This term is used for an electron tube that contains a gas or vapour, like mercury vapour, in sufficient quantity so that the electrical characteristics of the tube are determined entirely by the gas, once ionization has taken place. The *gas ratio* of this type of tube is the ratio of the ionization current in the tube to the electron current required to produce it.

Gas Multiplication : The production of additional ions by ions produced in a gas under the influence of a sufficient strong electric field is called gas multification.

It is also used as the ratio of the total ionization to the initial ionization as a result of the above process.

Gas Noise : The noise arising in a gas-filled tube due to the random ionization of the gas molecules in the tube. It is a form of white noise, *i.e.,* it has a relatively flat frequency spectrum; gas-discharge tubes are therefore used as white-noise generators for testing purposes.

Gassing : The evolution of gas in the form of small bubbles from one or more electrodes during electrolysis. Gassing occurs in an accumulator towards the end of the charging period.

Gate Expander : An array of diodes which are connected to the input stage of a diode-transistor logic gate and used to extend the possible number of inputs to that gate.

Gauss : Symbol Gs or G. It is defined as the unit of magnetic flux density in the obsolete CGS electromagnetic system of units. One gauss equals 10^{-4} tesla.

Gauss's Law for Electrostatics : This law relates to the electric field produced due to a stationary electric charge.

Geiger Counter Geiger Muller Counter : It is a gas filled tube used to detect ionizing radiation and count particles, in particular alpha particles. It consists of a thin wire anode mounted coaxially inside a cyclindeal cathode with a potential difference, slightly lower than that required to produce a discharge maintained across the electrodes.

The output characteristic of a Geiger counter shows a large plateau region where the output is constant for a different range of voltage; this is the operating region of the tube. The tube is also independent of the energy of the incident radiation in this region. Geiger counters are commonly used for count rates up to about 2000 counts per second.

Geissler Tube : It is regarded as a type of gas-discharge tube which gives the luminous effects of an electrical discharge through a gas at low pressure. It is used to investigate the spectra of gases and is also useful for spectrometer calibrations.

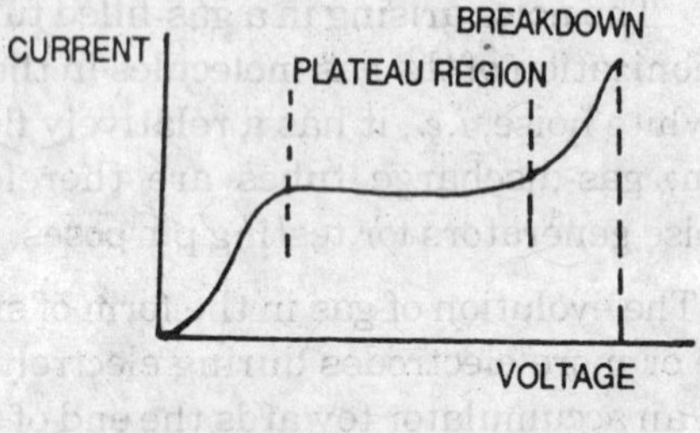

b-Current-voltage characteristic

Generator : This is defined as a machine that converts mechanical energy into electrical energy. An electromagnetic generator (*i.e., a* dynamo) has a coil which is made to move in such a manner as to cut lines of magnetic flux.

Germanium : A semiconducting element having symbol, Ge and atomic number 32, which find wide uses in semiconductor device. It has now been replaced by silicon as the substrate for most semiconductor devices.

Getter : This is a material which has a strong chemical affinity for other materials. This type of substances may be used to remove undesirable elements or compounds which are present in small quantities in an environment.

form as a *filament getter*, into a sealed vacuum system in order to remove residual gases. The process of using a getter is known as *gettering*.

Ghost : It is *syn.* to double image. It is defined as an undesired double image which appears on the screen of a television or radar receiver. It is due to ground-reflected waves arriving at the receiver slightly later than the direct waves.

Giga : It is a prefix to a unit, denoting a multiple of 10^9 of that unit: one gigahertz equals 10^9 hertz.

Gilbert : Symbol : Gb. It is the unit of magnetomotive force in the obsolete CGS electromagnetic system of units. One gilbert equals 0.7958 amperes.

Glass Electrode : It is defined as a thin-walled glass bulb containing a buffer solution into which dips a platinum wire, it means the wire is thus in electrical contact with the inner wall of the bulb. The device is used to measure the pH of a solution. The pH is determined by measuring the potential difference between the glass bulb (platinum contact) and the external solution.

Glass Valve : It is *syn.* to glass tube. It is defined as an electron tube having the envelope made entirely from glass without any different base material. The leads are taken in through special metal-glass seals.

Glow Lamp : It is defined as a gas-discharge tube operated in the glow-discharge region, the source of light is glow-discharge. The colour of the glow depends on the gas used in the tube : for example a *neon lamp*, containing neon, emits a red glow.

Goniometer : This is an instrument which measures angles, like the angle between two radiofrequency waves *(radiogoniometer)*. It is also used to find the direction of propagation of a wave.

Gramophone : Also called record player. This is an instrument that converts the undulations of the grooves of a gramophone record into sound. In this device stylus moves along the grooves and vibrates in sympathy with the modulations in the groove.

Graphical Symbols : These are the symbols that represent the various kinds of components and devices used in electronics, telecommunications, and allied subjects.

Graphic Instrument : It is a measuring instrument that presents the quantity being measured in the form of a graph. The graph is produced either in ink on a suitable paper chart or on the screen of a cathode-ray oscillograph.

Graphic Panel : It is defined as a master control panel in automatic and remote control systems in which coloured block diagrams are used to show the relation and functioning of the different parts of the control system.

Grass : The noise which takes place in radar receivers due to irregularities in the time base of the display. These may be as a result of random fluctuations in the time base generator or to electric interference. It results in noise on the picture which resembles grass.

Grassot Fluxmeter : This instrument is essentially a moving coil galvanometer in which both mechanical damping and restoring torque are made negligibly small. Moreover the Grassot fluxmeter has an advantage over the ballistic galvanometer for measuring the magnetic flux.

Gravity Cell : A type of primary cell which has two different electrolytes kept apart by their different densities.

Grenz Rays : The soft X-rays obtained if electrons are accelerated through 25 kilovolts or less.

Grid : It is defined as an electrode which has an open structure, like a mesh or a plate with a hole in it, permitting an electron beam to pass through it.

It is also regarded as the nationwide high-voltage transmission line system which interconnects many electricity power stations. It transmits voltages of up to 400 kilovolts.

Grid Bias : It is the potential applied to the grid of a thermionic valve which determines the portion of the characteristic curve at which the value will operate, or that modifies the cut-off value of the valve.

Grid Emission : The emission of electrons or ions from the grid of a thermionic valve is called grid emission.

Grid Leak : This is a high resistance connected between the grid and cathode of a thermionic valve which prevents an accumulation of charge on the grid.

Ground Absorption : It is defined as a loss of energy that occurs during transmission of radiowaves as a result of absorption in the ground.

Ground Plane : It is a sheet of conducting material that is adjacent to a circuit and is at earth potential. It can be used to provide a low-impedance earth at any point in the circuit.

Ground-Reflected Wave : It defined as a radiowave that travels between a transmitting and a receiving aerial situated above the ground and which undergoes at least one reflection from the ground.

Ground Reflection : This is the reflection of a transmitted radar wave by the ground before it reaches the target.

Ground Return : The echoes received by a radar receiver due to reflections of the radar wave by the ground or by objects on the ground.

Ground State : It is regarded as the lowest possible energy level of a system. In case a system has a configuration such that its energy is greater than its ground-state energy, it is said to be an *excited state*.

Ground Wave : It is a radiowave that travels between a transmitting and a receiving aerial situated above the earth. It has two main components : the space wave, which consists of two waves the direct wave and the ground-reflected wave, and the surface wave.

Grove Cell : It is a two-fluid primary cell in which a zinc rod immersed in dilute sulphuric acid acts as the negative element, which is separated from the positive element by a porous-partition.

Grown Junction : It is defined as a *p-n* junction which is formed in a single crystal of semiconductor material while the crystal is being grown from a melt. The amount and type of impurities added to the semiconductors are changed in a controlled manner as the crystal grows.

Guard Band : It is a frequency band which is left vacant during broadcasting to minimize mutual interference between two neighbouring frequency bands.

Guard Ring : It is a device used to give uniform fields and to define the sensitive volume in absolute electrometers and standard capacitors. It has a metal plate surrounding and coplanar with a smaller plate; a narrow air gap separates the two. The diagram shows the lines of force between a plate A surrounded by a guard ring RR, and a

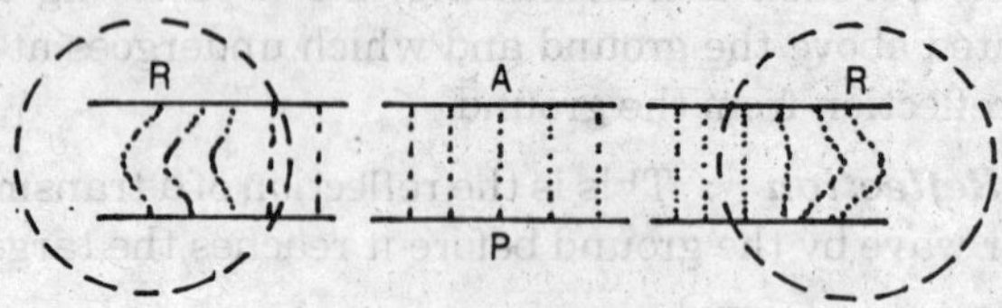

Principle of guard ring

parallel earthed plate P. The variations in field at the edges affect only the guard ring RR.

A similar device in the form of an auxiliary electrode is generally used in semiconductor devices and vacuum tubes.

Guard-Ring Capacitor : A standard capacitor which utilizes a guard ring to reduce the edge effect (see diagram). The *guard-well capacitor* is a special type of guard-ring capacitor applied for capacitance below 0.1 picofarads.

Gudden-Pohl Effect : It is regarded as the transient luminescence observed in a phosphor, previously exposed to ultraviolet radiation, in case an electric field is applied to it.

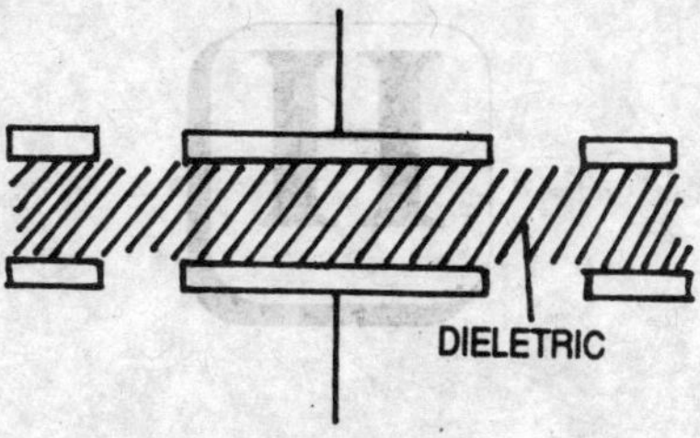

Guard-Ring Capacitor

Guillemin Line : It is network which gives pulses with very sharp rise and fall times in such a way that they are almost square.

Gunn Diode : A negative-resistance microwave oscillator which operates by means of the Gunn effect. It is a diode formed from a sample of low-resistivity n-type gallium arsenide which gives coherent microwave oscillations when a large electric field is applied across it.

Gunn Effect : It is an effect which takes place when a large d.c. electric field is applied across a short sample of n-type gallium arsenide. At values above the threshold value, typically several thousand volts per cm, coherent microwave oscillations are generated.

Gyrator : It is a component which does not obey the reciprocal theorem. In other words it reverses the phase of signals transmitted in one direction but does not affect signals transmitted in the opposite direction. Gyrators are generally employed at microwave frequencies, forming part of a waveguide.

H

Half-Power Point : It is defined as a point on a characteristic curve which corresponds to operation at a power intensity of half the maximum value.

Half-Wave Rectifier Circuit : It is a rectifier circuit which delivers unidirectional current to the load only during alternate half-waves of the singlephase alternating-current input.

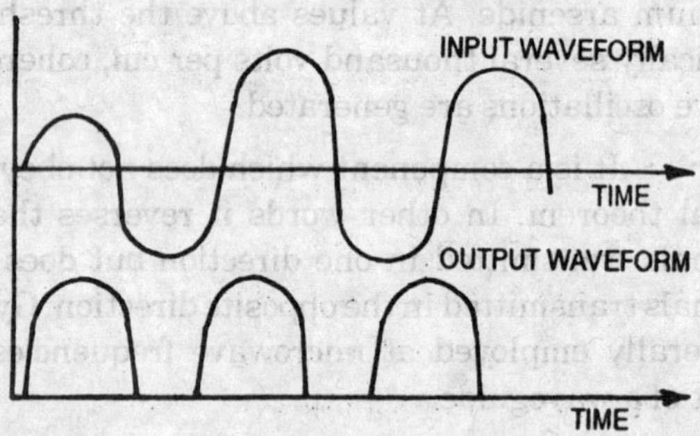

Half-wave rectification.

Half-Wave Voltage Doubler : It is a voltage doubler that operates only during the input wave half-cycle.

Hall Effect : This is an effect observed when a current-carrying conductor is placed in a magnetic field having direction perpendicular to both the direction of current and of magnetic field. E.H. Hall proved that charge carries in a metal conductor as electrons.

The force due to the magnetic flux density, B, on charge carriers of charge e and drift velocity v is equal to $ev \times B$. This force displaces the carriers and gives a nonuniform charge distribution giving rise to an electric field, E. At equilibrium the force on the charge carriers, eE as a result of this field just balances that due to the flux density given by the equation.

$$eE + ev \times B = 0$$

The drift velocity, v, is related to the current density, j, by the relation

$$j = nev$$

where n is the number of charge carriers per unit volume. In other words the electric field is related to the vector product of the magnetic flux density and the current density by the relation

$$E = -R_H\,(j \times B)$$

where R_H is the *Hall coefficient* and is equal to l/*ne*.

Hall-Effect Device : It is a device that utilizers the Hall effect in order to measure the charge carrier concentration in a metal or semiconductor or to detect magnetic fields.

Hall Probe : This is a method of measuring magnetic flux density, B, by using the Hall effect. A small sample of a suitable semiconductor, like indium antimonide or indium arsenide, is employed in which the Hall voltage developed across it varies almost linearly with the flux density.

Ham Radio : This is the non-commercial radio-communication between individuals. It should be noted that the frequencies used are restricted to internationally agreed values in order to prevent interference with commercial broadcast transmission and shipping or aircraft communication.

Hardware : It is defined as the physical units, devices, and circuits that make up a computer system.

Harmonic : It is an oscillation of a periodic quantity, present in a complex vibration, having a frequency which is an integral multiple of the fundamental frequency. An oscillation having a frequency which is an integral submultiple of the fundamental is termed as *subharmonic.*

Harmonic Analyser : It is regarded as a device that analyses a periodic function from its graph in terms of the Fourier series corresponding to the function.

Harmonic Generator : It is signal generator which generates harmonic of an input fundamental frequency. These devices may take the form of an oscillator controlled by a tuning fork.

Hartley Oscillator : This is a type of oscillator having a transistor with common-emitter connection and a parallel resonant circuit between the emitter and collector (*see* diagram). Ignoring the resistance of the coil the resonant frequency, ω_0 is represented by the equation :

$$\omega o^2 \sim C/L_1+L_2+2M)$$

Hartley oscillators may also be crystal-controlled.

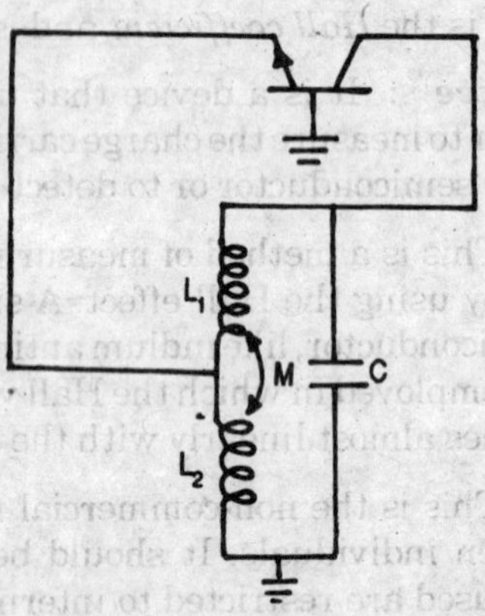

Hartley oscillator.

Hartshorn Bridge : It is an alternating-current bridge which measures mutual inductance. Direct comparison of a mutual inductance having a standard mutual inductance is virtually impossible because effects like the self and mutual capacitances of the coils, gives voltage components in the secondary circuits in phase with the primary current. This problem is solved in the Hartshorn bridge by the use of a variable resistor, of resistance r, common

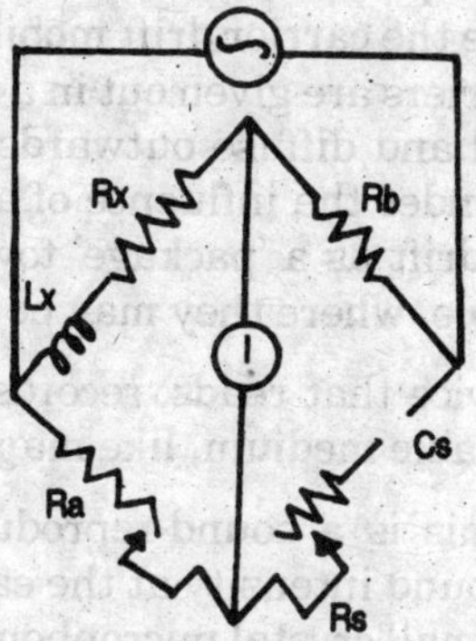

Hay Bridge

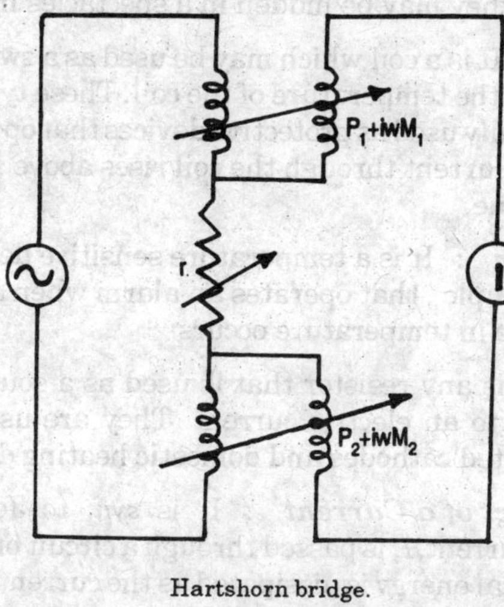

Hartshorn bridge.

to both the primary and secondary circuits of the bridge.

Hay Bridge : It is a four-arm bridge used for the measurement of large inductance (*see* diagram). At balance, when there is no response on the indicating instrument, I, we have

$$L_x = R_a R_b C_s / (1 + \omega^2 C_8^2 R_8^2)$$

$$R_x = L_x \omega^2 C_3 R_3$$

where ω is the angular frequency.

Haynes Schockley Experiment: : This is the experiment used to measure the carrier drift mobility in semiconductors. Excess carriers are given out in a sample of semiconductor material and diffuse outwards and recombine in the material. Under the influence of an external electric field they also drift as a 'package' towards the negative end of the sample, where they may be detected.

Head : It is a device that reads, records, or erases signals stored on a suitable medium, like magnetic tape.

Hearing Aid : This is a sound-reproduction system that increases the sound intensity at the ear. Modern hearing aids contain a small crystal microphone, battery-powered amplifier, and earpiece. The development of integrated circuits permit hearing aids to become very small and light and they may be hidden in a spectacles frame.

Heat Coil : It is a coil which may be used as a switch by the sensing of the temperature of the coil. These types of coils are generally used as protective devices that open a circuit when the current through the coil rises above a predetermined value.

Heat Detector : It is a temperature sensitive device, like a thermo-couple , that operates an alarm when a predetermined rise in temperature occurs.

Heater : It is any resistor that is used as a source of heat subjected to an electric current. They are used in indirectly heated cathodes and domestic heating devices.

Heating Effect of a Current : It is *syn.* to Joule effect. When a current, I, is passed through a circuit of resistance R, electrical energy is dissipated as the current flows from a higher to a lower potential. The rate of emission from the surface of the resistor just equals the rate of generation of heat.

Heating Time : It is the time taken by a device to heat up to a steady temperature.

Heat Sink : A device which is used to dispose of indirect heat in a circuit and prevent an excessive rise in temperature. These are particularly useful for protecting transistors in power applications.

Hecto : It is a prefix to a unit, denoting a multiple of 100 of that unit.

Height Control : It is a control in a television or radar receiver which adjusts the overall size of the frame scan.

Henry : It is the mutual inductance of two circuits when the current changing at the rate of one ampere in one circuit induces an *EMF* of 1 volt in other circuit.

Heptode : It is regarded as a thermionic valve that has seven electrodes, *i.e.*, five grids between the anode and cathode. The most common use of the device is as a mixer.

Hertz : Its symbol is Hz. The SI unit of frequency. It is the frequency of a periodic phenomenon that has a period of one second. It replaced the cycle per second (c.p.s)

Heterodyne : It is regarded as the production of beats in a radio receiver by the combination of the received wave with a locally generated wave of slightly different frequency.

Heterodyne Interference : It is a type of interference taking place in superheterodyne reception. It consists of a high audiofrequency note caused by beats between two carrier waves with almost equal frequencies.

Heterodyne Wavemeter : It is defined as device which is used to measure frequency. It consists of a variable-frequency oscillator whose output is combined with the frequency to be measured in such a way that beats are produced.

Heterojunction : It is a junction between two dissimilar semiconductors of opposite polarity types.

Heteropolar Generator : It is a type of electromagnetic generator, like a dynamo, in which the direction of the induced voltage in each conductor is reversed when that conductor successively traverses the magnetic field. In other words the direction of the magnetic field sensed by the conductor is reversed each time the conductor cuts the field.

Heusler Alloys : Alloys which show ferromagnetic properties although they contain no ferromagnetic material.

Hexode : It is defined as a thermionic valve having six electrodes, *i.e.*, four grids between the anode and cathode.

High-Density Packaging : It is the packaging of a given set of electrical components in a very small volume through the application of miniaturization techniques, as by the use of integrated circuits.

High Recombination Rate-Contact : It is defined as a contact between two semiconductors or between a semiconductor and a metal having a substantially constant charge carrier concentration whatever the current density at the contact.

High-Voltage Test : It is a test of the insulation of any electronic device. A voltage in excess of the normal operating voltage is applied across the insulation to ensure that it does not break down.

Hittorf's Principle : It states that if electrodes are arranged in a given pressure, discharge between them does not necessarily take place between the closest points of the electrodes. It will take place between points of potential difference equal to or greater than the threshold voltage required, although the threshold voltage depends on both the electrode separation and gas pressure.

H-Network : *Syns.* H-pad; H-section.

Hole : It is a term used for empty space appearing in the structure caused by leaving of an electron from the atoms from complex molecular structure of block of semiconductors material atoms linked to one another.

When an electric field is applied holes can drift through the material by a process of continuous exchange with adjacent electrons.

It should be noted that hole at or near the top of the valence band has positive effective mass and is mathematically equivalent to a positron.

Hole Conduction : It is the conduction of electricity in a semiconductor having a hole propagated through the crystal lattice under the influence of an electric field. An adjacent electron moves under the influence of the field

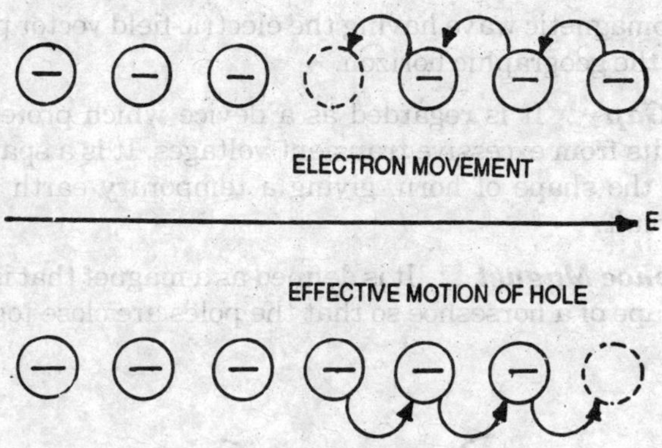

Hole Conduction.

and fills the vacancy, leaving a corresponding vacancy behind. The effective movement of the hole as a result of process of continuous exchange is equivalent to the movement of a positive charge in the same direction (*see* diagram). It should be noticed that holes move in the direction of positive field.

Hole Current : It is defined as the current in a semiconductor associated with the movement of holes through the material.

Hole Trap : It is a term used for a site in a semiconductor crystal, such as a donor impurity or lattice defect, that can trape holes.

Homeostasis : It is the condition of a system that is in dynamic equilibrium : the inputs and outputs are balanced in such a way that the system gives the appearance of a steady-state system.

Homopolar Generator : It is defined as an electromagnetic generator in which the voltage induced in the conductors always has the same sense with respect to the conductors.

Honeycomb Coil : It is a coil in which the turns are wound in a criss-cross fashion in order to diminish the distributed capacitance.

Hook-Up : It is a connection, generally of a temporary nature, between any electric or electronic circuits.

Horizontal Polarization : It is the polarization of an electromagnetic wave having the electric field vector parallel to the geographic horizon.

Horn Gap : It is regarded as a device which protects circuits from excessive transient voltages. It is a spark gap, in the shape of horn, giving a temporary earth for the circuit.

Horseshoe Magnet : It is defined as a magnet that is in the shape of a horseshoe so that the poles are close together.

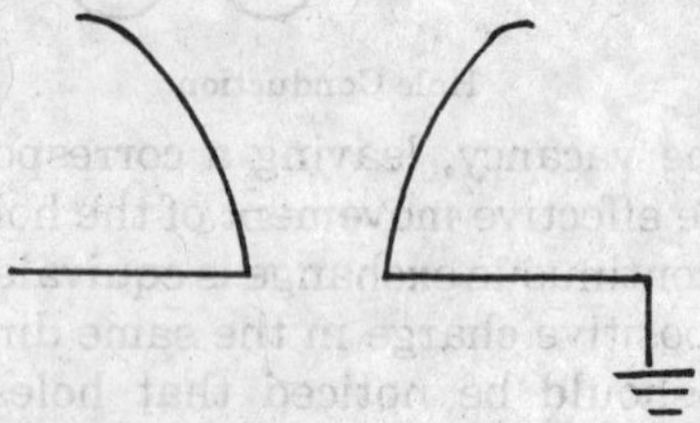

Home gap.

Hot-Cathode Stepping Tube : It is a type of multi-electrode thermionic valve with two or more stable paths that the electron beam can follow. A suitable input signal can result in the beam to step in a desired sequence from one position to another.

Hot Electron : A conduction electron in a solid which has an energy such that the electron is not in thermal equilibrium with the crystal lattice.

Hot Spot : A small portion of an electrode, like an arc spot which is at a higher temperature than the rest of the electrode.

Hot-Wire Instrument : A thermal instrument which utilizes the elongation by heat of a strip or wire carrying an electric current.

Hot-Wire Microphone : It is a device that utilizes the change in resistance of a hot wire in case it is exposed to sound waves. The change in resistance of the wire is equal to that due to a steady draught of velocity equal to the maximum velocity of the simple harmonic motion of the sound waves.

There are two factors contributing to the observed resistance drop : a steady drop depending on the particle velocity at the wire, and an oscillatory change due to the sinusoidal variations in amplitude of the sound wave.

Howler : It is defined as an oscillator that produces a high-pitched audio frequency tone in order to attract attention. It is used by telephone exchanges to attract attention to a telephone whose receiver has been left off.

It may a prose definial as an audiofrequency oscillator which is employed as a null-point detector in certain bridges. The oscillators give a high-pitched audiofrequency tone that either disappears or reaches a maximum intensity when the balance point is achieved.

HT. : *Abbrev. for* high tension.

Hum : It is a low-pitched audiofrequency droning noise heard in audiofrequency systems. It generally starts from the mains supply and takes place at frequencies that are harmonics of the mains supply frequency.

Hum Modulation : The unwanted modulation of a signal by hum arising in an audiofrequency system. It gives noise or distortion of the output.

Hunting : The fluctuation of a controlled signal about its desired value which is due to from overcorrection by a control device. The stable desired value is never actually attained.

Hybrid Junction : It is defined as a junction between transmission lines, either waveguides or coaxial lines. It should be noted that an ideal junction has no direct coupling between arms.

The two main types of hybrid junction are the *ring junction,* and the *T-junction.*

Hydrophone : It is a transducer which generates electrical signals in response to water-borne sound waves.

Hysteresis : The delay in the change of an observed effect in response to a change in the mechanism giving the effect.

The best-known example of hysteresis is magnetic hysteresis.

Hysteresis Distortion : The distortion introduced by non-linear hysteresis in one or more elements or a system.

Hysteresis Meter : It is an instrument that detects and measures magnetic hysteresis.

Hz : It is the *abbrev*. for hertz

I

I.C. : It is the *abbrev*. for integrated circuit.

Iconoscope : It is defined as a high electron-velocity photo-emissive camera tube having the target formed on a thin mica plate.

Ideal Crystal : It is defined as a crystal having perfect and infinite crystal structure and therefore contains no impurities or defects.

Ideal Transducer : It is defined as a transducer that produces the maximum possible signal power output under specified input and load conditions; i.e., the power loss in the transducer is negligible.

IF : It is *abbrev*. for intermediate frequency.

IF Strip : It is regarded as the part of the receiver in superheterodyne reception which is concerned with intermediate-frequency amplification. It consists of the intermediate-frequency amplifier together with its physical environment.

Ignition Coil : It is defined as such ignition coil that the voltage pulses output by the secondary coil are sufficiently large to produce sparks. The sparks may be used to ignite an inflammable mixture, such as the gases in an internal combustion engine.

Ignitor : It is an auxiliary electrode in an arc discharge tube which is employed to control the arc across the tube. It generally consists of a semiconducting rod with one end immersed in a mercury-pool cathode.

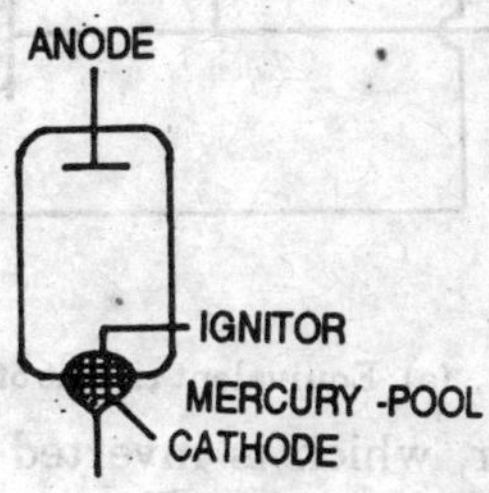

Ignitor

Ignitron : The ignitron consists of a cathode, anode and ignitor sealed in a tank. The ignitor is immersed in the mercury. Sending a pulse of current through the ignitor causes a small arc to form a point of contact of the ignitor and the mercury. It is this arc that initiates the anode current.

I^2L : *Abbrev. for* integrated injection logic. It is *syn.* to merged transistor logic (MTL). It is defined as a family of bipolar integrated logic circuits having very compact arrangement and provide a very high functional packing density combined with the possibility of high-speed operation. I^2L is essentially a development derived from diode transistor logic (DTL). The input diodes and transistors of DTL are combined in I^2L into a single multicollector *n-p-n* transistor and the current source is provided by a *p-n-p* transistor rather than a load resistor. It should be noted that the *p-n-p* transistor rather than a load resistor. It should be noted that the *p-n-p* transistor is formed in the same area of chip as parts of the multicollector transistor allowing a very contract structure. The multicollector structure gives the isolation between logic gates provided in DTL by the input diodes of the following gates.

The equivalent circuit is shown in (Fig. *a*). The *p-n-p* transistor is arranged laterally on the chip; the *p* type emitter is common to a large number of gates and is termed the *injector*. The *n-p-n* transistor is a vertical

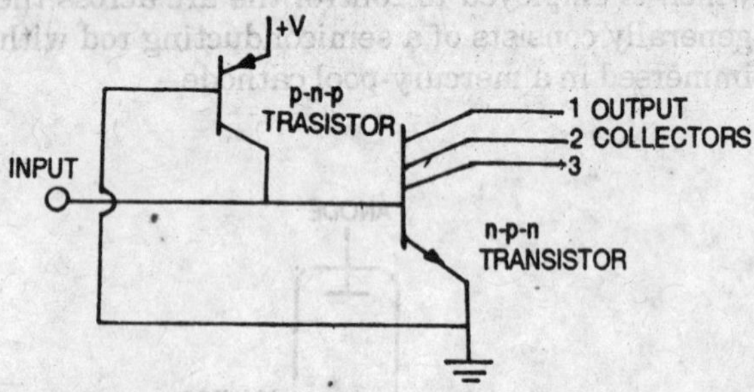

(a) Equivalent circuit of single I-L gate

transistor, which is inverted compared to the common method of fabrication. The *n*-type epitaxial layer forms the base of the *p-n-p* transistor and the emitter of the *n-p-n* transistor and is common to all the gates. An *n*+ guard ring surround cach *p*-type *p-n-p* collector/*n-p-n* base segion containing the *n*+ *n-p-n* collectors, in order to give isolation between individual gates. The fan-out is found by the number of multiple collectors.

Image Attenuation Constant : It is *syn*. to image attenuation coefficient.

Image Converter : It is *syn*. to image tube. It is defined as an electron tube that converts an image outside the visible part of the spectrum, such as an infrared image, into a visible image.

Image Dissector : It is a type of television camera tube, in which electrons from each portion of a photosensitive plate are focused in turn on a collector electrode in order to give a video signal.

Image Frequency : It is regarded as an unwanted input frequency that arises from a source other than that to which a receiver used in superheterodyne or beat reception is turned and that causes superious signals to be output.

Image Orthicon : It is defined as a low-electron-velocity camera tube in which light from a scene in focused on a photocathode having a lightsensitive material deposited on a thin sheet of glass. Electrons are emitted from the photocathode in proportion to the intensity of the light and are focused on to a target having a thin glass disc with a fine mesh on the photocathode side of the disc. The impact of electrons from the photocathode results in secondary emission of electrons from the target which is greater than, but proportional to, the original electron density from the photocathode. The secondary electrons are collected by the mesh screen and drained off to a power supply. It should be noted that the target is left with a positive static charge pattern corresponding to the original light image.

Image Phase Constant : It is *syn.* to image phase-change coefficient.

Image Potential : It is defined as the potential at a point just outside the surface of a material due to an electric image.

Image Transfer Constant : Its symbol is θ. It is defined as a complex quantity, given by $(\alpha + i\beta)$ of a quardipole network terminated in its image impedances.

The real part of the image transfer constant (α) gives the image attenuation constant and the imaginary part (β) gives the image phase constant.

Impact Ionization : It is the ionization of an atom or molecule as a result of the loss of orbital electrons following a high-energy collision. In a semiconductor electron-hole pairs can be produced if the electrons have sufficient energy to enter the conduction band.

IMAPTT Diode : The term is used for a semiconductor diode which acts as a powerful source of microwave power. When a *p-n* junction is reverse-biased into avalanche breakdown, it exhibits negative resistance at microwave frequencies and may be used as a negative-resistance oscillator.

Impedance : Its symbol is Z. It is the total opposition to the passage of current is an *a-c* circuit and like reactance is measured in ohm. The impedance, is given by the ratio of the voltage to the current in the circuit. The alternating current is given by the relation

$$I=I_o \cos \omega t$$

where I_o is the peak current and ω is the angular frequency. In case reactance, due to the capacitance and inductance, is present in a circuit, the voltage will be out of phase with the current and given by the equation

$$V=V_0 \cos(\omega t+\phi)$$

where ϕ is the phase angle and is given by

$$\phi=\tan^{-1}[(\omega L-1/\omega C)/R]$$

where R is the resistance, C the capacitance, and L the inductance in the circuit. The voltage is represented by the equation $V_0\cos(\omega t + \phi) = IR + L(dI/dt) + 1/C\int I dt$

Solving this equation gives that the current is equal to

$$I=V \cos \omega t/\sqrt{[R^2+(\omega L-1/\omega C)^2]}$$

The impedance, which is the ratio of current to voltage, is a complex quantity as I and C are phasor quantities. The magnitude of Z is given by

$$|Z| =\sqrt{[R^2=(\omega L-1/\omega C)^2]}$$

$$=\sqrt{(R^2=X^2}]$$

where X is the reactance.

Impedance Matching : It is regarded as the matching of the impedances of the parts of a system to ensure optimum conditions for transfer of power from one part of the system to another.

It should be noted that the maximum power is transferred from an amplifier to a load when the load impedance is made the conjugate impedance of the amplifier output impedance.

Impedor : It is *syn*. to impedance. It is defined as a circuit element, like a resistor, capacitor, or inductance, that is used mainly for its impedance.

Impregnated Cable : It is a cable with paper insulation in which the insulating properties of the paper tapes are improved by covering them with an insulating compound.

Impregnated Carbon : It is defined as the carbon in an arc lamp which is mixed with other materials in order to produce a coloured arc.

Impregnated Cathode : It is defined as a cathode consisting of porous tungsten impregnated with a barium salt in such a way that as the barium evaporates from the surface of the cathode it is continuously replenished from the bulk of the material.

Impulse Generator : It is *syn*. to surge generator. It is defined as an electronic device which produces a single pulse, typically one representative of the surges generated by lightning in a transmission line.

Impulse Voltage : It is defined as a unidirectional voltage which gives a maximum value and then falls to zero more or less rapidly without any appreciable superimposed oscillations. An impulse current is a unidirection current with similar characteristics. Impulses are generally undesired, being generated by fault conditions in electrical equipment or apparatus or by procedures such as switching on or off.

A typical waveshape is depicted in Fig. a. The maximum voltage, V, of the impulse is the peak value. It should be noted that the *wavefront* is the rising portion of the OA and the wavetail is the falling portion ABC. The duration of the wavefront is the time, T_1 required for the voltage to rise from zero to the peak value. The time to half value of the wavetail, T_2, is the time required for the impulse to rise from zero, reach the peak value, and then decay to half the peak value. The waveshape in Fig.a is illustrated as a T_1/T_2 impulse. The average rate of increase of voltage with time is termed as the steepness of the wavefront. A typical impulse voltage which is due to lightning is a /50 wave, where T_1 and T_2 are measured in microseconds; standard testing equipment uses a wave with these proportions produced by a surge generator.

An impulse voltage or current frequently results in flash-over or puncture in electrical apparatus. When this takes place the impulse voltage collapses rapidly and is described as a *chopped impulse voltage* (Fig. *b*). Flashover or puncture can take place either during the wavefront or during the wavetail. The actual value of the impulse voltage at which flashover or puncture take place during the wavefront when is the *impulse flashover voltage* or the *impulse puncture voltage*. When flashover or puncture occurs on the wavetail the peak value of the impulse voltage is quoted. The time to flashover or the time to puncture is regarded as the time interval between the beginning of the impulse voltage or current and the instant at which the wave is chopped (point *D* in Fig. *b*).

Impurities : These are the foreign atoms in a semiconductor which are either naturally existing or deliberately introduced. Impurities have a fundamental effect on the type and amount of conductivity of the semiconductor.

Incandescence : It is defined as the emission of visible light from a substance at a high temperature. The term is also applicable to describe the radiation itself.

Incandescent Lamp : The term in used for an electric lamp in which light is produced by incandescence. In case an electric current is passed through a suitable filament of carbon, osmium, tantalum, or tungsten the temperature of the filament is raised sufficiently for incandescence to take place. Temperatures in excess of 2600°C are produced and a noble gas filling is generally used in the lamp to prevent disintegration of the filament.

Incremental Plotter : It is defined as a device which gives a graphic display from data generated by a computer. A smooth curved line is not generally achieved due to the digital nature of the computer output; small increments in the *x*-, *y*-, and diagonal directions are used to approximate the curves.

It should be noted that an incremental plotter is generally used as an off-line plotter, the source of data is a previously prepared magnetic tape, disk, or paper tape and the plotter may be located remotely from the computer. An incremental plotter may be linked directly to the computer and used as an on-line plotter. In other words the points are plotted graphically as soon as they are generated.

A *microfilm plotter* is an incremental plotter which gives a graph on photographic film rather than paper.

Incremental Recorder : It is off-line device that records data, as it is generated, in a form suitable for inputing directly to computer. Suitable recording media include magnetic tape, paper tape, and punched cards.

Index Error : This is an error on the scale of a measuring instrument in such a way that under zero input conditions a reading x is registered by the instrument.

Indicator Tube : An extremely small cathode-ray tube, having screen diameter in millimetres. It is used to show the value of a varying quantity by altering the size or shape of the image on the screen in accordance with the input voltage.

Indirect Wave : This is regarded as the portion of a transmitted wave which does not travel directly from the transmitter to the receiver but is reflected by the ionosphere.

Induce : It is to cause an electrical or magnetic effect in a circuit or device by altering the condition of a seccnd circuit or body.

Induced Current It is a current which flows in a conductor due to changing magnetic flux density, in which the lines or magnetic flux intersect with the conductor.

Induced Noise : It is defined as a noise which appears in any circuit, device, or apparatus due to electromagnetic induction from other nearby circuits.

Inductance : Its unit is henry. It is as defined a constant that relates the magnetic flux, ϕ linking a circuit to the current flowing in the circuit or in a nearby circuit.

The *self-inductance* (symbol : L) of a circuit is defined as one henry if the circuit is threaded by a total flux of one weber when a current of one ampere is flowing. The *mutual inductance* (symbol : M or L_{12}) between two circuits is defined as one henry a flux of one weber is present in one circuit due to a current of one ampere flowing in the second circuit.

Induction Coil : It is defined as a device that uses electromagnetic induction to generate a series of pulses of high potential and approximately unidirectional current. It has a primary coil of wire with only a few turns wound on an iron core, surrounded by a coaxial secondary coil of many turns and insulated from it.

Induction Compass : It is a device that indicates direction. It consists of a small coil that is made to rotate in the earth's magnetic field. The direction indicated by the compass depends on the magnitude of the induced current in the coil.

Induction Flowmeter : It is device which measures the rate of flow of a conducting liquid. The liquid is made to flow through a tube of length L, placed in a magnetic field of flux density B (see diagram). Electrodes are placed on a diameter of the tube and the induced e.m.f., E, across this diameter depends on the rate of flow, v, of the liquid: and is given by the equation

$$E = v \times BL$$

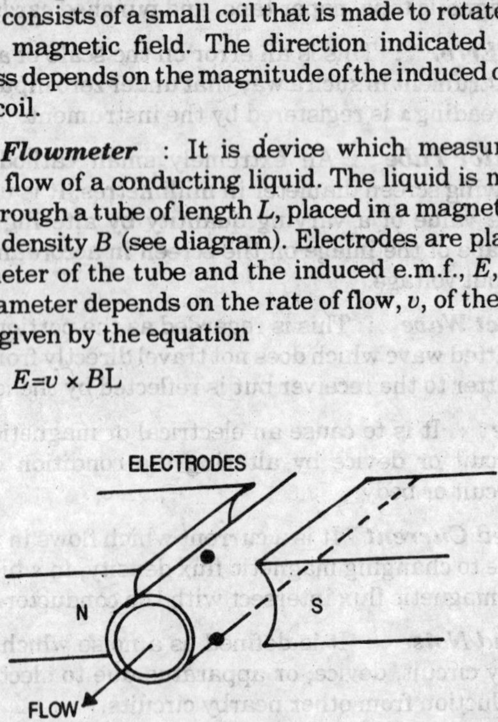

Induction flowmeter

Induction Furnace : It is a device which employs the heating effect of an induced current in the secondary of a transformer.

Induction Generator : It is an alternating-current (a.c.) generator having a coil or number of coils which are made to rotate in an alternating magnetic field produced by a number of electromagnets excited from an alternating-current source. The necessary condition is that the rotating coils are driven at a sufficient speed–determined by the a.c. source frequency and the number of pairs of magnetic poles–a.c. power is supplied to a load at the same frequency as the exciting voltage.

Induction Heating : The heating which is due to induced eddy currents in a conducting material when subjected to a varying magnetic field.

Induction Instrument : It is defined as an instrument which depends for its operation on the interaction between a magnetic field which is changing in a fixed winding and the eddy currents induced by the field in a movable conductor.

Induction Machine : It is a machine that generates electric charge by electrostatic induction.

Induction Microphone : It is a microphone having a straight-line conductor as the moving element.

Induction Motor : It is an alternating-current motor in which the current in the secondary winding is generated by electro-magnetic induction when alternating current is supplied to the primary winding generally, the stator. Mechanical movement is produced from the torque by the interaction between the rotor current and the magnetic field due to the current in the stature.

Inductive Interference : The interference in a communication system as a result of electromagnetic induction from the electric supply system.

Inductive-output Device : It is a device in which the output voltage is generated by electromagnetic induction between the current and the output electrode, without the current carriers being collected by the output electrode.

Inductor : It is *syn*. to inductance; reactance coil. It is defined as a device or circuit element, generally in the form of a coil, which has inductance and is used primarily because of that property.

Inert Cell : It is regarded as a primary cell containing the chemicals and other necessary ingredients in solid form, but will not function until water is added to form an electrolyte.

Inertia Switch : It is a switch which operates when an abrupt change in its velocity takes place.

Information Theory : It is an analytical technique which gives optimum amount of information required to transmit a message or solve a specific problem in communication, control, or computer systems.

Infrared Radiation : It is defined as a portion of the electromagnetic spectrum of radiation extending from the limit of red end of the visible spectrum to the microwave region. The infrared region is in the region of 730 nanometeres to one millimeter in wavelength.

Inhibiting Input : It is an input applied to a digital gate which prevents any output that might otherwise occur.

Injection : In general the term is referred to the application of signal to an electronic circuit or device.

It is also used for the introduction of excess charge carriers, either electrons or holes, into a semiconductor material so that the total number present exceeds that at thermal equilibrium.

Injection Efficiency : It is defined as the ratio of the current carried by injected minority carriers to the total current across the junction in the case of *p-n* junction

Input : It is the signal or driving force applied to a circuit device computer, machine, or other plant.

It is also referred to apply as an input signal or driving force.

Input Impedence : It is the impedence of a circuit or device, presented at its input.

Input/Output : It denotes all those operations, devices, and data bearing media which are employed to pass information into or out of a computer.

Insertion Loss : The loss of power in a load which takes place when a network in inserted between the load and the generator supplying the load is called insertion loss.

Instantaneous Carrying-current : It is defined as the peak value of current which may be carried instantaneously by a switch, circuit-breaker, or similar device at rated voltage under specific condition.

Instantaneous Frequency : It is defined as the rate of change of phase of any oscillating electric variable, measured in radians per second divided by 2 π.

Instantaneous Power : The value of the rate at which power is transmitted from that circuit to the next portion of the system at the output terminals of a circuit.

Instantaneous Value : It is defined as the value, at a particular moment, of any quantity that varies with time.

Instrument Rating : It is regarded as the limits, designated by the manufacturer, between which an instrument can operate without damage.

Instrument Sensitivity : It is defined as the response of a measuring instrument expressed as the ratio of the magnitude of the physical response to the magnitude of the quantity measured. It is usually directly mentioned as for example, divisions per volt.

Instrument Transformer : It is a transformer used in conjunction with a measuring instrument. It utilizes the *current transformer* or *voltage transformer* properties of a transformer.

Insulate : It is a term used to support or surround a conductor with insulating material so that the current through it is confined to desired path.

Insulated System : It is a system of distribution of electric energy which is not earthed under normal conditions.

Insulating Barrier : It is a type of screen of insulating meterial which is fitted to certain electrical apparatus, such as switch gear or fuses.

Insulating Resistance : It is defined as resistance between two electrical conductors of systems of conductors which are normally insulated from each other. It is generally of the order of megohms but in the case of cables it is expressed as megohms per kilometre.

Insulation : Material which is employed to insulate an electrical conductor. It is also the method of insulating a conductor.

Insulator : It is defined as any form of matter in which free electrons either are not available or are present in exceedingly small quantities. An insulator is used to prevent the loss of electric current or charge from a conductor.

Integrated Circuit (I.C.) : It is defined as a complete circuit which is manufactured as a single package. A *hybrid integrated circuit* has several separate component parts attached to a ceramic substrate and interconnected either by wire bonds or a suitable metallization pattern.

Integrating Frequency Meter : It is *syn.* to master frequency meter. It is defined as instrument that allows a check to be made on a source of alternating voltage by integrating the number of cycles which take place in a specified time interval. Comparison with the number of cycles which should take place in the same time interval reveals whether the prescribed frequency has been maintained.

Integrator : This is a simple R.C. network with a long time constant. Output is taken across the capacitor. In practice integrator is a summing device.

A *capacitance integrator* utilizes a capacitor, usually in series with a resistor, to perform integration.

Intelligibility : It is a function of a speech communication system like a telephone system, in which it is more important for the received information to be intelligible than for the input to be flawlessly reproduced.

Intensity Modulation : It is defined as the variation in brilliance of the spot on the screen of a cathode-ray tube in accordance with the magnitude of an input signal.

Interaction Space : It is a region in an electron tube that roughly corresponds to the interelectrode space and in which the electrons interact with an alternating magnetic field.

Interactive : Interactive operation enables a user at a far off place to send and receive information to and from a computer quickly, and to modify the operation of a program during its execution following the production of intermediate results or interrogation.

Interconnection : It is defined as any method of providing an electrical path between any of the materials which combine to form a circuit.

These are connections between and external to any functional item which forms a circuit or system of circuits. Functional items mean component parts, devices, subassemblies, and assemblies.

Interelectrode Capacitance : The capacitance between specified electrodes of an electronic device which can form a small capacitor within the device. The operation of these type of devices can be significantly affected by the existence of interelectrode capacitances.

Interference : It is regarded as disturbance to the signal in any communication system obtained as a result of undesired signals.

Interlock : It is a safety device which permits a piece of apparatus to function only when predetermined conditions are fulfilled.

Internal Resistance : Symbol : r, unit: ohm. It is defined as a small resistance possessed by a cell, accumulator, or dynamo. It is represented by the equation.

$$r=(E - V)/I$$

where E is the e.m.f. generated by the device, V is the potential difference across the terminals, and I is the current.

International Ampere : Its symbol is A_{int}. It is the former standard of electric current defined as the constant current which when passed through an aqueous solution of silver nitrate, will deposit silver at the rate of 0.001 11800 grams per second, it was replaced as the standard unit by the ampere in year 1948

1 A_{int}=0.99985 abamperes

1 A_{int}=9.9985 amperes

International System : It is a former system of units

which expressed the values of electrical quantities in terms of the international ampere, international ohm, centimetre, and second. It has now been replaced by SI units.

Interrupter : It is defined as a device, like an induction coil, that periodically interrupts a continuous current.

Interstage Coupling : The coupling between successive stages of a multistage amplifier employing several amplifying stage in cascade is called interstage coupling.

Intraconnection : It is an electrical connection inseparably associated with circuit elements within a component part.

Intrinsic Conductivity : It is regarded as the conductivity of a semiconductor associated with the semiconductor material itself and not contributed by impurities.

Intrinsic Crystal : It is *syn.* to iodichromatic crystal. It is defined as a crystal that exhibits photoelectric properties in its pure state.

Intrinsic Mobility : The mobility of charge carriers in an intrinsic semiconductor is called intrinsic mobility. The mobility of electrons in approximately three times that of holes.

Intrinsic Semiconductor : It is a pure semiconductor which contains equal concentrations of electrons and holes under conditions equal concentrations of electrons and holes under conditions of thermal equilibrium. Absolute purity is unobtainable in practice and thus nearly pure materials are termed intrinsic.

Intrinsic Temperature Range : It is the temperature range in which the electrical properties of a semiconductor are not necessarily modified by impurities in the crystal.

Inversion : It is the production of a layer of opposite polarity at the surface of a semiconductor under the influence of an electric field, generally an applied one. Spontaneous inversion can take place in the surface of p-type material when is in contact with an insulating layer due to the presence of positive ions in the insulator. Inversion can only take place when sufficient mobile minority carriers are present in the semiconductor material, otherwise a depletion layer forms.

Inverter : It is a device which changes direct current into alternating current. A common type of inverter is a rotating machine designed for the purpose.

It is *syn.* to digital inverter; NOT circuit. *it* is also defined as a logic circuit that inverts the level of the input signal, in other words which produces a low output for a high input and vice versa.

Inverting Transistor : It is a bipolar or MOS transistor used as an analong or digital inverter.

Ion : It is defined as an atom with either a positive or negative electric charge. Usually each atom has an equal number of electrons and protons. The atom is therefore electrically neutral; ions are created by adding or removing electrons; the number of protons in an atom is constant.

Ion Burn : It is defined as a defect of the phosphor coating on the screen of a cathode ray-tube that is caused by bombardment by heavy negative ions present in the beam.

Ion Density : It is defined as the number of pairs of positive and negative ions per unit volume.

Ionic Atmosphere. : The atmosphere, consisting of ions which surrounds an individual ion in an electrolyte is called ionic atmosphere.

In other wards the symmetry of the ionic atmosphere is disturbed with respect to the surrounded ion (*see* diagram) and the ions experience a retarding force as a result of the migrations of the opposite-polarity ionic atmosphere in the opposite direction.

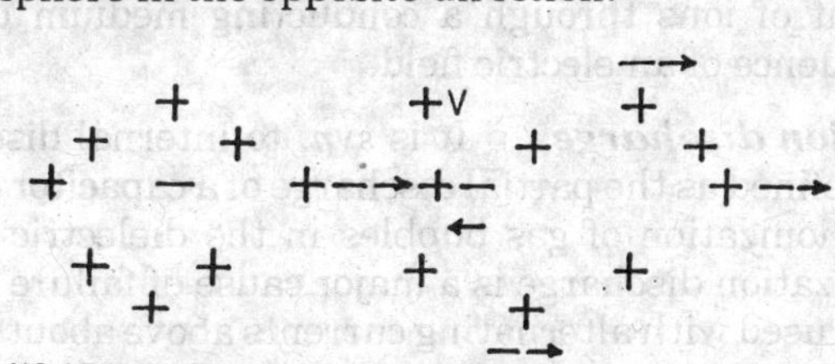

Ionic atmosphere

Ionic Conduction : It is defined as the movement of changes through a semiconductor as a result of the displacement of ions within the crystal lattice, such movement being maintained by a continuous supply of external energy.

Ionic Crystal : It is *syn.* to electrovalent crystal. It is defined as a crystal that is in the form of an array of positively and negatively charged ions, the interatomic forces being of the Coulomb type.

Ionic Heating : The heating of a surface by ion bombardment.

Ionic Mobility : It is velocity of an ion when moving under the influence of unit electric field.

Ionic Semiconductor : It refers to a semiconductor in which the movement of ions through the material contributes more to the conductivity than the movement of mobile charge carriers.

Ion Implantation : It is regarded as the technique of implanting ions into the lattice of a semiconductor crystal by bombarding the surface with ions under controlled conditions.

Ionization : Any method in which ions are formed form neutral atoms. Ions are formed readily when an electrolyte dissolves in a suitable solvent.

Ionization Chamber : It is a gas-filled radiation-detection tube which is employed to detect ionizing radiation. The ionization chamber is a very sensitive radiation detector and can be used to detect and measure a wide range of energies and intensities of radiation.

Ionization Current. : The current which is due to movement of ions through a conducting medium under the influence of an electric field.

Ionization discharge : It is *syn.* to internal discharge. It is defined as the partial discharge of a capacitor caused by the ionization of gas bubbles in the dielectric material. Ionization discharge is a major cause of failure of capacitors used with alternating currents above about 200 volts.

Ionization Gauge : It is a pressure gauge which is employed to measure extremely low gas pressures down to

10^{-8} mmHg. It has high-vacuum three-electrode thermionic valve fused to the gas system (*see* diagram). Electrons leaving the cathode are accelerated to the grid but fail to reach the anode due to its negative potential. Any gas molecules in the system are ionized by collision with some of the electrons which pass through the grid; the positive ions thus formed migrate to the anode generating an output current which is a function of the number of gas molecule present and thus of the gas pressure.

Ionization Potential : It is *syn.* to electron binding energy; radiation potential. Symbol: I; unit: electronvolt. It is the minimum energy required to remove an electron from a given atom or molecule to infinity.

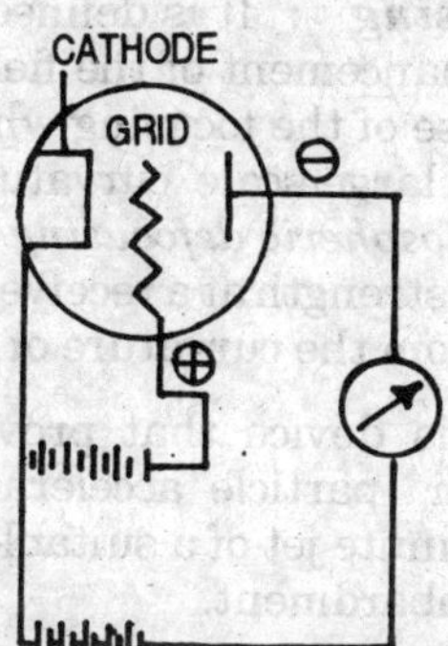

Ionization gauge

The electron generally removed from the atom or molecule is in the outermost orbit, i.e., the least strongly bound electron. Some atoms and molecules may be ionized by the removal of an electron which is not in the outermost orbit, i.e., one that is more strongly bound. The resulting ion will be in an excited state. The ionization potential for removal of the least strongly bound electron is termed as the *first ionization potential* (I_1) and results in an ion in the ground state. Ionization by the removal of the second (or subsequent) least strongly bound electron requires a greater ionization potential, termed the second (or third, etc.) ionization potential (I_2, I_3, etc.).

Ionizing Event : Any process which gives an ion or group of ions. The physical agent involved in the process, like a charged particle passing through a gas, is termed as *ionizing agent.*

Ionizing Radiation : Any radiation which produces ionization or excitation of the medium through which it passes.

Ionosphere : It is a spherical shell of ionized air which surrounds the earth, extending from about 50 km to over 1000 km, and is employed to reflect radiowaves in order to obtain long-distance radio transmission between points on the earth's surfance.

The ionosphere consists of several distinct layers or regions which may change in thickness between day and night and also show seasonal and latitude variations.

Ionospheric Focusing : It is defined as a process that results in an enhancement of the field strength at a given receiver because of the focusing which arises from either small-scale or large-scale curvature of the ionospheric layers. The *ionospheric defocusing* results from a reduction of the field strength at a receiver due to the defocusing which arises from the curvature of the layers.

Ion Source : It is device that provides ions, which are mostly used in particle accelerators. A common type consists of a minute jet of a suitable gas that is subjected to electron bombardment.

Ion Spot : It is a defined as a darker area appearing on the screen of a cathode-ray tube and resulting from decreased luminescence of the screen (not necessarily of the phosphor) after bombardment by heavy negative ions in the beam.

It is also regarded as a spurious signal appearing in the output of camera tubes and image converters due to alteration of the charge pattern on the target or cathode by ion bombardment.

Ion Trap : It is regarded as a device used in cathode-ray tubes which attracts heavy ions present in the electron beam and thereby prevents them impinging on the phosphor coating of the screen and causing blemishes.

Iron : It is a metal having symbol Fe and atomic number 26, which exhibits ferromagetism and has a high tensile strength. It is mostly employed in electronics as a magnetic material and for screening purposes. It is relatively abundant and therefore has a low cost.

Irradiation : It is defined as the exposure of a body or substance to electromagnetic or corpuscular ionizing radiation.

Island Effect : It is the effect, occurring in thermionic valves when the grid voltage falls below a certain minimum value, in which the emission from the cathode is restricted to a few small areas of the cathode.

Isochronous Cyclotron : It is a form of cyclotron which takes account of the relativistic mass change of the accelerated particles. Above a particular energy (about 15 MeV/proton) the relativistic mass change results in orbital time (Be/mc) of the particles to fall out of synchronism with the accelerating alternating voltage.

Isoelectronic : This term denotes a groups of atoms which show similar electronic properties, such as similar distributions of electrons in the outer orbits.

Isolating : It is a term used for disconnecting a circuit or device from an electric supply system, generally by making a circuit open at a time when it carries no current.

Isolating Transformer : It is a transformer that is used to isolate any circuit or device from its power supply; thus the circuit derives power from the source without a continuous wire connection between them.

Isolation Diode : It is defined as a diode used in a circuit to allow signals to pass in one direction but to block those in the other direction and therefore prevent damage from surges in the reverse direction.

Isolator : It is a device, generally made of a ferrite, which permits microwave energy to pass in one direction with little loss but absorbs power in the reverse direction.

Isotopes : There are two or more nuclides that have the same atomic number but differ in nuclear mass due to different numbers of neutrons in the nucleus. Each nuclide is said to be an isotope of the element defined by the given atomic number.

Isotropic : It is denoting a substance in which physical properties, such as magnetic susceptibility, are the same in all directions.

Iterative Impedance : It is defined as the impedance that when connected to one pair of terminals of a quadripole produces a like impedance at the other paid or terminals. In general a quadripole has two iterative impedances, one for each pair of terminals.

J

Jack Plug and Socket : It is regarded as a kind of type of plug-and-soket connector which is employed if rapid and easy connections between circuits or devices are required. Insertion or removal of the plug may result in one or more switching functions to take place, such as the breaking of a short circuit. The basic construction is depicted in the diagram. The plug slides into the sprung socket and is correctly located by means of a groove running round the plug. It should be noted that the contacts are arranged

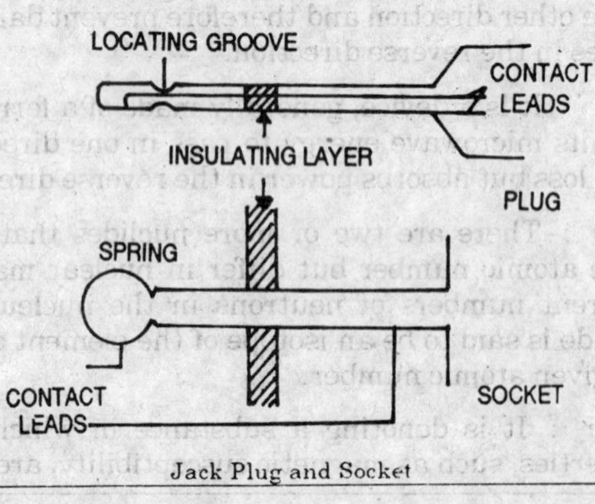

Jack Plug and Socket

linearly along the length of the plug and socket and are insulated from each other. Two or more contacts can be used. The end of the plug forms one contact and the others lie along the length of the metal casing. Contact is obtained by means of a central wire or conducting rod. More than one rod may be employed when more than two contacts are required, the rods being insulated from the casing and from each other except at the connection point.

Jamming : The deliberate interference in communications and radar by means of undesired signals which are intended to render unintelligible or to falsify the whole or part of the desired signal. A *jammer* is mostly employed to produce the jamming signal.

Jitter : The short-term instabilities in either the amplitude or phase of signal. It causes momentary displacements of the image on the screen, giving it a shaky or 'jittery' appearance. An oscilloscope that is used to measure the amount of jitter is known as a *jitter scope*. Momentary errors of synchronization between the scanner and receiver in television or facsimile transmission may result in jitter of the received images. This type of jitter is known as *jitters*.

Johnson- Lark-Harowitz Effect : It is defined as the change in resistivity of a metal or a degenerate semiconductor as a result of scattering of the charge carriers by impurity stoms.

Johnson-Rahbeck Effect : It is regarded as an effect observed when a high potential difference of about 200 volts is applied across two plates, one of metal and one of a semiconductor material, such as slate or agate. The two plates are held together as a result of the applied voltage.

Josephsom Effect : It is defined as an effect which takes place when a sufficiently thin layer of insulating material is introduced into a superconducting material. A superconducting current may flow across the junction, known as a *Josephson junction*, in the absence of an applied voltage. This is termed as the *direct current Josephson effect*. In case the value of the current exceeds a critical value. I_c, determined by the properties of the insulating barrier, current can only flow when a unite voltage is

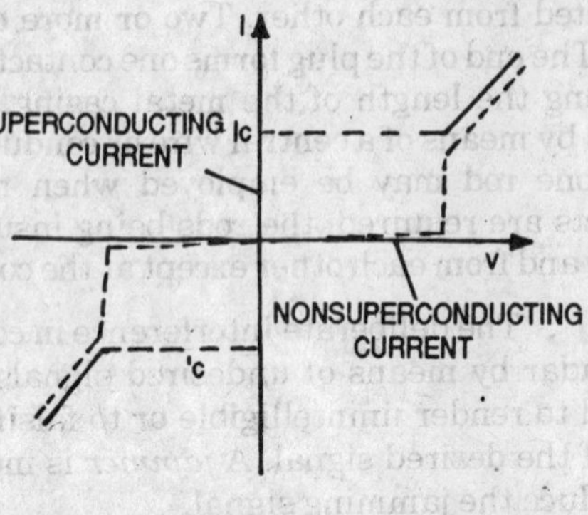

I-V characteristic of Josephson (unction)

applied. The current voltage characteristic is shown in the diagram, in which the dashed curve is the current-voltage characteristic in the nonsuperconducting state.

Josephson Memory : It is defined as a cryogenic memory which has an array of Josephson cells, i.e. memory cells containing Josephson junctions, held at a temperature very close to critical temperature.

Joshi Effect : The variations of the current in a gas discharge due to the effect of light on the gas.

Joule : Its symbol is J. It is the SI unit of energy, including work and quantity of heat. It is defined as the work done when the point of application of a force of one newton is displaced one metre in the direction of the force. One joule equals one watt-second.

JUGFET : It is *abbrev. for* junction field-effect transistor.

Jump : It is a transition of an orbital electron from one atomic energy level to another.

Jumper : It is a direct electrical connection between two points in a printed circuit that is not a part of the interconnection pattern of the circuit.

Junction : It is defined as a contact between two different conducting materials such as two metals, as in a rectifier or thermocouple.

It is also defined as a connection formed between two or more conductors of the same type or between sections of transmission lines.

Junction Box : It is an enclosed distribution panel for the semipermanent connection or branching of electrical circuits.

Junction Coupling : The coupling of cavity resonator to a coaxial line by direct connection to the coaxial conductor.

Junction Transistor : It consists of three sections, the two end sections of the same type *P* or *N* and the central section of the other *N* or *P*. It means that either *P* is between two *N*-type or *N* is in between two *P* type sections.

N Type. Pure gemanium is semiconductor when small traces of impurity like antimony is added to it, the conductance is slightly increased. The electrons are free to move in the crystal and this type of crystal is called *N*-type.

P Type. When an impurity like gallium is added in germanium, holes or gaps are created in the germanium crystal because the impurity attracts electrons from the crystal. Therefore the crystal is deficient in electrons and results the holes having the charges. These holes can migrate throughout the crystal.

K

K Band : It is defined as a band of microwave frequencies ranging from 10.9 to 36.0 gigahertz.

Keeper : It is defined as a piece of ferromagnetic material, generally soft iron, which is placed across the extremities of one or more permanent magnets when not in use (see diagram). The keeper completes the magnetic circuit and neutralizes the demagnetizing effect in the magnet.

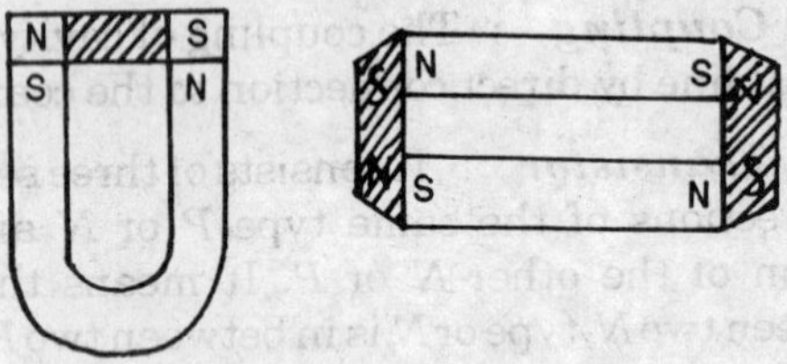

Keepers

Kelvin : Its symbol is *K*. It is defined as a the basic SI unit of thermodynamic temperature, defined as 1/273.16 of the thermodynamic temperature of the triple point of water, i.e., the equilibrium point at which pure ice, air-free water, and water vapour can coexist in a sealed vacuum flask. It is related to the *degree Celsius* (°C) by the relation

$$i°C = T - 273.15 \text{ kelvin}$$

where T is the thermodynamic temperature and 273.15 K is the temperature of the ice point, which is the zero of the Celsius (or centigrade) temperature scale.

Kelvin Balance : It is a type of instrument in which the electromagnetic forces resulting from the passage of a current through a set of coils are balanced against the force of gravity..

Kelvin Contacts : It is a means for testing or making measurements in electronic circuits or components, particularly when low values are being measured.

Kelvin Double Bridge : It is *syn.* to Thomson bridge. This is a bridge with six arms which has been developed from the Wheatstone bridge for measuring low resistances. The bridge is connected as shown in the diagram X is the low resistance to be measured and S is a standard low resistance. Two sets of variable resistances, R_1, R_2 and r_1, r_2 are varied together until balance is achieved. At balance we have:

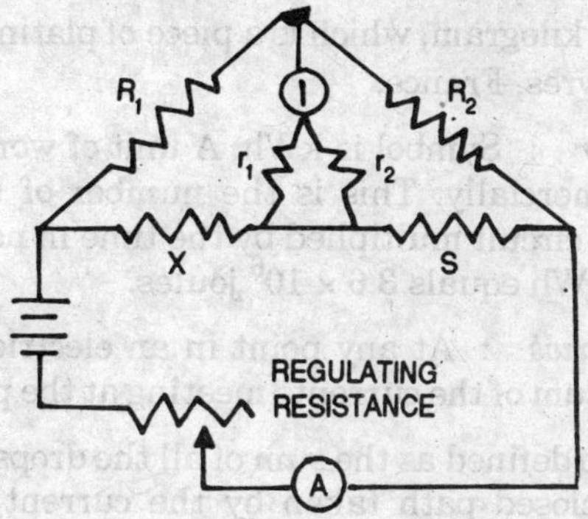

Kelvin Double Bridge

$$R_1/R_2=r_1/r_2=X/S$$

Kelvin-Varley Slide : It is defined as a device used in vernier potentiometers to diminish the effect of contact resistance. The device consists essentially of two or more sets of slide wires or resistance coils in cascade, each coil acting a a decade voltage divider for the preceding one.

Kerr Effects : The two effects in which the optical properties of transparent material are affected by electric or magnetic fields.

The *electro-optical effect* is the effect by which the direction of polarization of plane-polarized light through a refractive medium is rotated by an electric field applied perpendicularly to the direction of propagation of the light.

The *magneto-optical effect* occurs when plane-polarized light is reflected from a highly polished pole face of a strong electromagnet. Slight elliptical polarization of the light beam is produced.

Kilo : Its symbol is *K*. It is a prefix to a unit, denoting a multiple of 10^3 (i.e., 1000) of that unit: one kilometre equals 10^3 metres.

It is also as a prefix in computing to denote a multiple of 2^{10}(i.e., 1024): one kilobyte equal 2^{10}bytes.

Kilogram (or Kilogramme) : Its symbol: kg. It is the SI unit of mass equal to the mass of the international prototype of the kilogram, which is a piece of platinum-iridium kept at Sevres, France.

Kilowatt Hour : Symbol is kWh. A unit of work or energy used commercially. This is the number of kilowatts of power in a circuit multiplied by the time in names it is in use. One kWh equals 3.6×10^6 joules.

Kirchhoff's Laws : At any point in an electric circuit the algebraic sum of the currents meeting at the point is zero.

It is also defined as the sum of all the drops around the complete closed path taken by the current is equal to applied e.m.f.

Kirk Effect : It is an effect observed in epitaxial power transistors designed for microwave applications. At high injection levels the current is sufficiently high so that the injected minority carrier concentration in the base becomes comparable to or greater than the base doping concentration.

Klydonograph : It is defined as a type of surge-voltage recorder which has a photographic film or plate placed between two electrodes.

Klystron : It is defined as an electron tube that is used as a microwave amplifier or oscillator. It is a linear-beam microwave tube having velocity modulation which is applied to an electron beam in order to produce amplification of a microwave-frequency field.

The most important type of klystron is the *reflex klystron,* which is employed as a low-power oscillator. This type of klystron has only one cavity, which acts as both buncher and catcher. Velocity modulation of the electron beam is caused by the input radiofrequency wave in the cavity, and the modulated beam leaving the cavity is reflected back by a *reflector electrode*. Bunching takes place as a result of the faster electrons travel further towards the reflector before reversing their direction of travel than do the slower ones. The bunches of electrons returning to the cavity which experience the maximum positive field give up the most energy as the direction of motion is now reversed.

Multicavity klystrons are employed when either extremely high power pulses are required, as in the power source for a particle accelerator, or of continuous waves of moderate power are needed, as in ultrahigh-frequency television transmitters. Three or more resonant cavities coupled to the electron beam are employed to provide a high overall gain.

Knife Switch : It is a type of switch which has fixed contacts and movable current-carrying blades. The blades are hinged in such a way that on operation of the switch each blade moves in its own place to make or break contact.

Koch Resistance : It is defined as the resistance of a photocell in case light is incident on the active surface of the tube.

L

Lag : It is defined in the following ways: (i) The amount, determined as a time interval or the angle in electric degrees, by which one periodically varying wave is delayed in phase with respect to the similar phase in another wave.

(ii) The time interval between the transmission of a signal and its detection by a receiver.

(iii) The delay between a correcting signal of a control system and the response to it.

(iv) The persistance of the electrical image in a television camera tube; it may be several frames in duration.

Lagging Current : It is regarded as an alternating current that has a lag with respect to the applied electromotive force producing it.

Lagging Load or Inductive Load : It is a reactive load in which the inductive reactance exceeds the capacitive reactance and thus carries a lagging current with respect to the voltage across the terminals.

Lamination : It is a thin stamping of iron or steel, oxidized or lightly varnished on the surface, which is used in the assembly of a laminated core for use in transformers, transductors, relays, chokes, or similar apparatus.

Land : It is a metallic contact area in microminiature electronic devices or circuits.

Langevin Ion : It is the ion that moves through a gas under the influence of an electric field:

Laplace Operator Symbol : ∇^2 The differential operator is

$$\partial^2/\partial x^2 + \partial^2/\partial y^2 + \partial^2/\partial z^2$$

Laryngophone : It is *syn.* for throat microphone.

Laser : A source of intense monochromatic coherent radiation in the visible, ultraviolet, or infrared regions of the electromagnetic spectrum. It should be noted that the narrow beam can be either pulsed or continuous. A laser operates on similar principles to the maser but the transitions between the energy levels are in a higher energy range.

Lateral Recording : A method of recording a gramophone record or magnetic disk in which the groove modulations are parallel to the surface of the recording medium and perpendicular to the motion of the cutter.

Lattice Constant : A parameter which gives the configuration of a crystal lattice. The lattice constant is represented by the lengths of the edges of a unit cell of a crystal or as the angle between the axes of the cell.

Lattice Parameter : It is *syn.* to lattice spacing

Lawnmower : *Colloq.* A type of preamplifier employed with radar receivers that reduces the level of grass on the screen.

L Band : A band of microwave frequencies ranging from 0.39 to 1.55 gigahertz.

LCD : It is used as *abbrev. for* liquid-crystal display.

L-C Network : A turned circuit having both inductance and capacitance. The product, *LC*, of inductance and capacitance is constant for any given frequency.

Lead : It is an electrical conductor, which is used to make external connections between circuits or pieces of apparatus. It is used in the form of a wire or cable.

It is the amount, measured as a time interval or as the angle in electric degrees, by which one periodically varying wave is advanced in phase with respect to the similar phase in another wave.

Lead Symbol: Pb.: : It is a metal having atomic number 82, and is mainly used either as an alloy, with other metals like tin, to form solders or in storage batteries.

Lead-Acid Cell : A type of accumulator having sponge lead cathodes and lead dioxide anodes and employs dilute sulphuric acid as the electrolyte.

Lead-In : The cable connecting the active part of an aerial to the transmitter or receiver.

Leading Current : An alternating current having a leads with respect to the applied electromotive force producing it.

Leading Load or Capacitive Load : A reactive load generally containing resistance and capacitance which carries a leading current with respect to the voltage across the terminals. It should be noted that a pure capacitance introduces a lead of 90^0 or one quarter wavelength.

Leakage : The passage of an electric current along a path other than that intended because of faulty insulation or isolation in a circuit, component, device.

Leakege Current : A fauit current taking place in any electronic device circuit, etc., due to leakage. It is small compared with a short-circuit current.

Leakage Flux : Lost flux in any apparatus, like a transformer, which has a magnetic circuit. The leakage flux is flux which is outside the useful portion of the flux circuit.

Leakage Indicator : An instrument which detects and measures the value of leakage current to earth in any electrical system.

Leakage Reactance : It is the undesired reactance in a transformer or alternator due to leakage flux cutting one coil but not the other. A leakage inductance is given out by this effect and leads to losses in the system.

Lecher Line : A transmission line having a short section of two parallel open wires which may be used for the measurement of wavelength or for impedance matching.

Leclanche Cell : A primary cell having a carbon-rod anode and a zinc cathode, which may be amalgamated. The electrolyte is 10-20 per cent ammonium chloride solution.

Leduc Effect or Righi Effect : When heat flows through a strip of conductor or semiconductor placed in an orthogonal magnetic flux density, a temperature gradient is obtained across the strip at right angles to both the direction of heat flow and the magnetic flux density.

Lenard Spiral : It is a small noninductively wound spiral of bismuth wire mounted between mica plates and is used to measure magnetic fields. The change in resistance due to a magnetic field is found by connecting the spiral to one arm of a Wheat-stone bridge.

Level Compensator : A device or circuit which automatically compensates for the effects of amplitude variations in a received signal.

Leyden Jar : It is an early type of capacitor.

Life Test : It is a test in which a sample or population of items is subjected to stated stress conditions for specified times with stated failure or success criteria, in order to determine its reliability characteristics.

Lifetime : It refers to the mean time interval between generation and recombination of a charge carrier in a semiconductor.

Light-emitting Diode : (LED) A *p-n* junction diode which gives out light as a result of direct radiative recombination of excess electron-hole pairs-the Lossev effect-which is one of the recombination processes in a semiconductor.

Lightning Conductor : It refers to a lightning protective system having a single conductor providing a path between an air terminal and earth along which a lightning stroke can pass.

Lightning Stroke : It is regarded as an electric discharge due to the discharge of one of the charged regions of a thunder cloud. The polarity of the lightning stroke is the polarity of the electric charge which comes to earth. It should be noted that the complete *lightning fiash* is a complete discharge along a single path.

Light-pen : A pen-like device which gives data, in the form of a visible image on the screen of a cathode ray tube, by 'writing' in a manner similar to that with a pen on paper. These are used in conjunction with an on-line visual display unit which inputs the data produced to a computer. It means that they are extremely powerful design tools.

Limited Stability : It refers to the property of any system, circuit, device, etc. which is stable only for a specific range of values of input signal and unstable outside this range.

Limiter : Any device that automatically sets a boundary value or values upon a signal. The term is generally used for a device which, for inputs below a specified instantaneous value, gives an output proportional to the input, but for inputs above that value gives a constant peak output.

Linear : It refers to any system, device, or apparatus that has its essential physical parts arranged in a line, as in a linear accelerator.

Linear Accelerator : A particle accelerator using electrons or protons which are accelerated as they travel along a straight evacuated chamber. The acceleration is provided by the electric field vector associated with the radio-frequency (r.f.) output from a klystron or magnetron.

Linear Circuit or Analog Circuit : It means a circuit in which the output varies continuously as a given function of the input.

Linearly Graded Junction : It refers to a junction between two different-polarity semiconductors (*p-n*, *p-i*, or *n-i*) having the concentration of impurities varying linearly across the junction.

Linear Scan : It refers to a sweep of the electron beam in a cathode ray tube in which the beam scans the screen with constant velocity, generally, by application of a sawtooth waveform to the deflection plates or coils.

It also refers to a scan using a radar beam that moves with constant angular velocity.

Linear Time-base Oscillator : A relaxation oscillator which is employed to produce generator as sawtooth waveform for use as a time base.

Line Choking Coil or Screening Reactor : An inductive reactor which may be connected in series with electrical plant, like a power transformer, in order to protect such plant from sudden steep-fornted or high-frequency surges in the power supply.

Line Communication : It refers to the communication, between two points by means of a physical path like wire or waveguide.

Line of Flux : It is an imaginary line drawn in a magnetic field whose direction at any point along its length is that of the magnetic flux density, B.

Line of Force : It is an imaginary line drawn in an electric field whose direction at any point along its length represents that of the field at that point.

Line Printer : An output device used with computers and data-processing system which prints an entire line of characters at a time rather than a character at a time.

Line-sequential Colour Television : It refers to a colour-television system in which each of the video signals (red, blue and green) is transmitted in turn for the duration of one entire scanning line.

Link : It refers to a communication channel or circuit which is employed to connect other channels or circuits.

It also refers to a path between two switches that form part of a central control system in automatic switching.

Lin-log Receiver : A type of radar receiver which has a linear output for small input signals and a logarithmic output for large ones.

Liquid Crystal : An organic liquid consisting of long-chain molecules which line up under the influence of an external electric field to give a quasi-crystalline structure to the liquid.

Liquid crystals are temperature-sensitive.

Liquid crystal Display (LCD) : A type of passive display which employs liquid crystals, as in the seven-segment numerical display of digital watches and pocket calculators.

Liquid rheostat : A variable resistor containing the resistance element which is a liquid metal.

Lissajous Figure : It refers to the displacement pattern traced out when two sinusoidally varying quantities are superimposed at right angles to each other. These patterns may be obtained on the screen of a cathode-ray

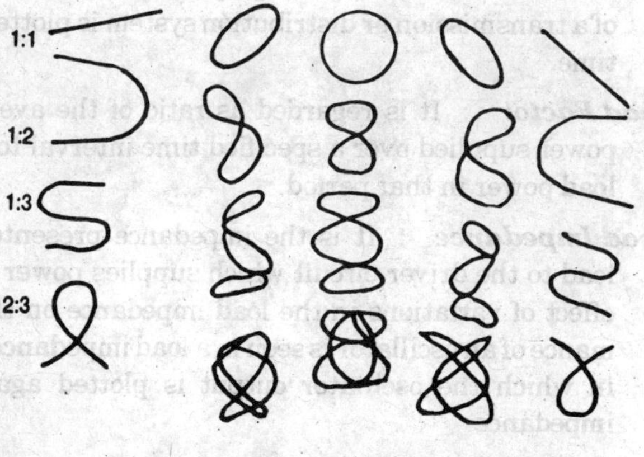

Lissajous Figure

oscilloscope by using the two signals to the horizontal and vertical deflection plates.

The simplest pattern is a straight line, which takes place when two signals of equal frequency and in phase with each other are superimposed. Some complicated figures result when the ratio of frequencies is not a simple one. Typical figures are shown in the diagram for various frequency ratios and phase angels.

Litzendraht Wire : A multistranded wire formed from many fine conducting filaments.

Load : A load is the point where useful work is being done e.g., electrical signals televisions etc.

Load Characteristic : A characteristic curve for an electronic device, like a transistor, showing the dynamic relationship between the instantaneous values of two variables if all the supply voltages are maintained constant. Emitter current, I_e, plotted against emitter voltage, V_e, provides an example.

Load Circuit : It means the output circuit of any electrical machine, generator, transducer, circuit, or device.

Load circuit Efficiency : It is regarded as the ratio of the useful power delivered to the load by the circuit of any device, to the power input to the device.

Load Curve : It refers to a graph in which the load power of a transmission or distribution system is plotted against time.

Load Factor : It is regarded as ratio of the average load power supplied over a specified time interval to the peak load power in that period.

Load Impedance : It is the impedance presented by the load to the driver circuit which supplies power to it. The effect of variations in the load impedance on the performance of an oscillator is seen in a load impedance diagram in which the oscillator output is plotted against load impedance.

Load Leads : It refers to the conductors or transmission line which connect the power source for dielectric or induction heaters to the load, load coil, or applicator used.

Load Line : It is a line drawn on the graph of a family of characteristics of an electronic device, like a transistor, which shows the relationship between current and voltage of the circuit under consideration for a given load.

Load Matching : It refers to adjustment to the output impedance of the load circuit of a dielectric or induction heater to optimize the energy transferred from the source to the load.

Load Regulator : It is a circuit or device which maintains a load at a constant value or varies it in a predetermined way.

Load Transfer Switch : It s a switch which is employed to connect the power output from a generator or other power source to one or to other load circuits as desired.

Lobe Swiching or Beam Switching : It is a kind of scanning used in radar in which the direction of maximum radiation or reception is switched sequentially to each of a number of preferred directions.

Lock-in Detector or Lock-in Amplifier : It refers to detector that responds only to an input signal whose frequency is synchronous with the frequency of a locally generated control signal. It is generally employed as a null-point detector in a bridge circuit.

Locking : It refers to controlling the frequencies of an oscillator with the help of an applied signal of constant frequency from an external source.

It also refers to latching which means holding a circuit in position or in a certain state until previous operating circuits are ready to change the circuit.

Locking-in : It means synchronizing the frequencies of two coupled oscillators, as in a frequency doubler, so that the two frequencies have a desired ratio, usually of two integral numbers.

Locking-on : It means following of a target, automatically, by a radar aerial.

Logarithmic Resistor : A form of variable resistor designed in such a manner that the fractional change of resistance is directly or indirectly proportional to the movement of the contact.

Logical One : The digit 1 used in binary notation. It is equivalent to the value 'true' of a logical statement. *Logical zero* is the digit 0 in binary notation. It is equivalent to the value 'false'.

Logic Circuit : It refers to a circuit designed to perform a particular logical function based on the concepts of 'and', 'either-or', 'neither-nor', etc. Mostly these circuits operate between two discrete voltage levels, *i.e. high* and *low logic levels,* and are said to be *binary logic circuits.* Logic using three or more logic levels is possible but not so common.

Logic Diagram : A diagram which gives the logic elements of a computer or data-processing system, or a function thereof, and their interconnections but does not generally show any constructional or engineering details. A logic diagram is very helpful when designing such systems or smaller networks to perform a specific mathematical operation like integration.

Logic Element : It is regarded as the smallest circuit blocks in a computer or data-processing system that may be represented by mathematical operators in symbolic logic.

Logic Symbol : A graphical symbol which represents an element.

Long-Line Effect : This effect is sometimes observed when an oscillator is coupled to a load through a transmission line which is long compared with the wavelength of the oscillator output.

Long Persistence Screen : This is a screen which is employed in cathode-ray tubes which permits the image on the screen to persist for several seconds.

Long Tailed Pair : These refer to a pair of matched bipolar transistors having their emitters coupled together with a common emitter bias resistor acting as a constant current source. The name is derived from the bias resistor that physically resembles a tail.

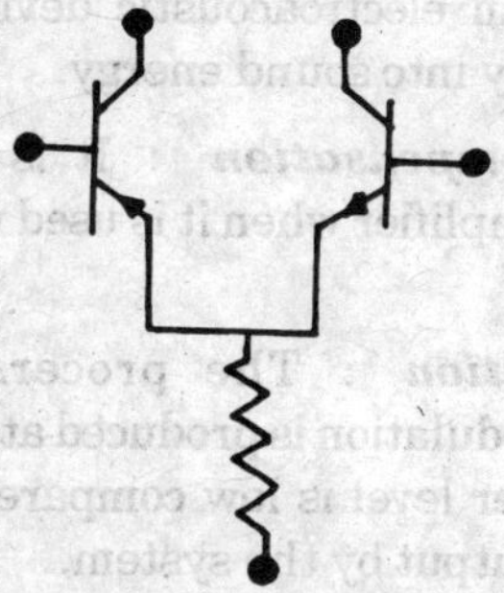

Long Tailed Pair:

Long Wave : A radiowave having a wavelength in the range on to 10 km.

Loop Aerial or Frame Aerial or Coil Aerial : It is a kind of directive aerial having high value of directivity that consists essentially of a coil with one or more turns of wire of very small axial length compared to the diameter.

Lorentz Force : It is the force experienced by an electron or other charged particle moving in a region of magnetic flux density.

Loss : A circuit is said to have loss when the reginal power out put is less than the signal power input.

Loss Angle : It is he angle by which the angle of lead of the current is less than 90^0 when a capacitor or dielectric is subjected to sinusoidal alternating electric stress.

Loss Factor : It is regarded as the ratio of the average power dissipation in a line, circuit, or device to the power dissipated at peak load.

Lossless Line : A hypothetical transmission line in which no attenuation takes place.

Lossy : It is an insulator which dissipates more energy than is considered normal for the class of material.

Lossy Line : A type of transmission line which is designed to give a high degree of attenuation.

Loudspeaker : An electroacoustic device which converts electrical energy into sound energy.

Low-Frequency Compensation : It is the compensation applied to an amplifier when it is used with low-frequency signals.

Low-Level Modulation : The procers of modulation whereby the modulation is produced at a stage in a system where the power level is low compared with the level of power that is output by the system.

Low-Loss Line : It refers to a kind of transmission line which is designed to dissipate very little energy per unit length.

Luminescence : It is the emission of electromagnetic radiation from a substance due to a nonthermal process. The term is also employed to described the radiation itself.

Lumped Parameter : Any circuit parameter, like inductance, capacitance, which can be treated as a single parameter at a point in the circuit for the purposes of circuit analysis over a specified range of frequencies.

M

Magic Eye : It refers to cathode-ray tuning indicator; electron indicator tube; electron-ray tube. A miniature cathode-ray tube generally used in radio receiver and as an indicator for voltmeters and galvanometers in a.c. bridges. It shows a flourescent pattern which is evaluated by the rectified output voltage of the receiver, and is an aid to tuning the receiver or finding the null point in a.c. bridges.

Magnesium : Symbol: Mg. A light metal having atomic number 12, used extensively for the construction of electronic components, mostly as an alloy with aluminium.

Magnet : A body having the property of magnetism. The term is applied to those bodies which can give an appreciable magnetic field external to themselves. Magnets are either temporary or permanent.

Magnetic Amplifier : An amplifier containing transductors which are used to produce power amplification of the input signal.

Magnetic Balance : An instrument which determines directly the force between two magnetic poles.

It is used to measure the strength of a magnetic field.

Magnetic Bubble Memory : It means a serial magnetic memory which is fabricated as a solid-state device and in which information is stored as microscopic domains of magnetic polarization. The memory has a nonmagnetic garnet substrate-G^3- on which an epitaxial layer of magnetic garnet (usually calcium germanium YIG) is grown. In the presence of a sufficiently large steady magnetic flux density applied perpendicular to the surface, the direction of magnetic polarization aligns in the direction. Any small domains of opposite polarization, which can be produced (or destroyed) by local variation of the magnetic flux density, are stable and may be moved through the surface using suitable weaker magnetic flux densities applied parallel to the surface.

The small domains are called *magnetic bubbles* and are obtained by electromagnetic induction using a single coil on the surface. They are caused to move through the surface by means of a periodic structure of suitably shaped magnetic electrodes obtained from a different material, such as permalloy. A rotating magnetic flux density parallel to the surface makes the electrodes alternately to attract and repel the bubbles.

Magnetic Circuit : A completely closed path attributed to a given set of lines of magnetic flux.

Magnetic Contactor : It is a contactor which is operated by magnetic means, like an alternating magnetic field.

Magnetic Crossed Field Modulator : It is a type of frequency doubler having an annular coil surrounded by a hollow toroid of magnetic material. The toroid act as a core of an external toroidal winding. The toroidal winding carries a direct voltage and the alternating current is input to the annular winding. The output voltage developed in the toroidal winding gives a frequency twice that of the input signal due to the interaction of the orthogonal magnetic fields in the coils.

Magnetic Damping : It is the process of damping in an indicating instrument containing a damper made of a metal vane, connected to the pointer, that moves through a magnetic field.

Magnetic Field : The space surrounding a magnet or a current carrying conductor and containing collection of flux line around. It may be represented by lines of force having direction at any point in the direction of the force exerted on a small coil (a search coil) placed in the field at that point. The direction of the force is normal to the magnetic flux density at that point. It should be noted that the dimensions of the coil are sufficiently small so as not to disturb the magnetic conditions.

Magnetic Field Strength : *Or magnetizing force* Symbol H; Unit ampere/metre. The strength of a magnetic field at a point in the direction of the line of force at that point. It is defined in vacuo by the equation.

$$B=\mu o H$$

where B is the magnetic flux density and mm is a constant, the permeability of free space.

Magnetic-Fip-Flop : It is a flip-flop which employs one or more magnetic amplifiers as the bistable element.

Magnetic Flux : Symbol : b; unit: weber. It is the flux through any area in the medium surrounding a magnet or current-carrying conductor, equal to the surface integral of the magnetic flux density over the area. It is determined by the e.m.f. produced when a circuit linking the flux is removed from it. One weber of flux linking a circuit of one turn produces an e.m.f. of one volt in that circuit when the flux is reduced to zero.

Magnetic Flux Density or Magnetic Induction : Symbol: B; unit: tesla 9weber/metre2). The fundamental force vector in magnetism, the magnetic analogue of the electric field E.

It means that the unit of magnetic flux density is one which exerts a force of one newton on a wire of length one metre carrying a current of one ampere.

Magnetic Hysteresis : It is a phenomenon observed in ferromagnetic materials below the Curie point where the magnetization of the material changes nonlinearly with the magnetic field strength and also lags behind it. The magnetic susceptibility of these materials is large and positive, a large value of magnetization *M* being produced for comparatively small fields. A characteristic plot of their magnetization. M, or magnetic flux density, B, against magnetic fields strength, *H,* indicates the hysteresis effect and is termed a *hysteresis loop.*

The effect of hysteresis in a magnetic core results in an

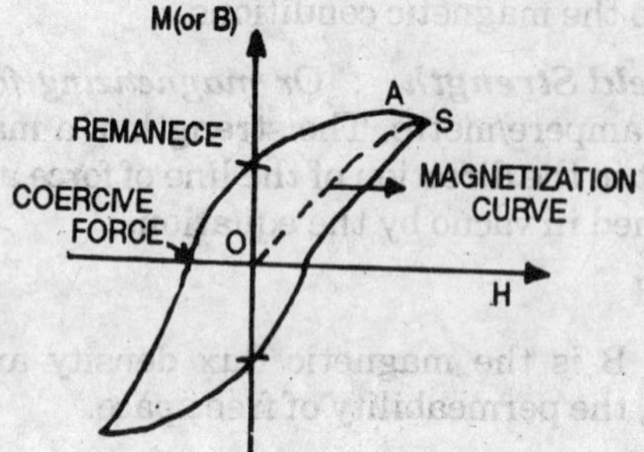

Magnetic Hysteresis loop.

increase in the effective resistance of the coil surrounding the core; the *hysteresis factor* is the increase in the effective resistance of a coil carrying a current of one ampere at a specified frequency.

Magnetic Leakage : The loss of magnetic flux from a magnetic circuit as a result of a portion of the total magnetic flux following a path that makes it ineffective for the desired function of the circuit. It should be noted that magnetic leakage reduces the overall efficiency of the operation.

Magnetic Memory : It refers to a memory having digital information represented by the direction of magnetization of magnetic material.

Magnetic Moment or Magnetic Dipole Moment

: Symbol m; unit: ampere metre^2. A measure of the strength of a magnet. When a magnet is placed in a homogeneous magnetic flux of magnetic flux density B it experiences a torque T gives by the equation where m is the magnetic moment. When the

$$T = m x B$$

magnet is a small coil of area dA carrying a current I, then the magnetic moment m is equal of IdA.

Magnetic Pole : It is the region of concentrated magnetism in a magnet analogous to an electrostatic point charge.

Magnetic Recording : It is the process of recording electrical signals on a magnetic medium to produce a reasonably permanent easily reproducible record which is flexible in use and produces an accurate reproduction of the original signal.

The best known application of magnetic recording is sound recording an *magnetic tape* known as *tape recording*. In magnetic sound recording the magnetic tape is moved at a uniform speed past the poles of an electromagnetic and in this way is longitudinally magnetized. It should be noted that during reproduction the process is reversed: the tape is fed past an electromagnet (the *replay head*) and the changes of magnetization induce currents in the coils corresponding to the original magnetizing currents.

The recording medium is generally made from finely devided ferrous oxide particles deposited on a plastic backing known as the base. The base is cellulose acetate tape 0.25 inches wide.

The electromagnet which is employed to record, reproduce, or erase the signal on the tape is termed as a *head (or magnetic head)*. It is possible to employ a single *read-write head* to perform each separate function but commercial tape recorders generally employ separate heads for recording, reproducing, and erasing. A typical head consists of soft iron pole pieces wound with wire coils. The

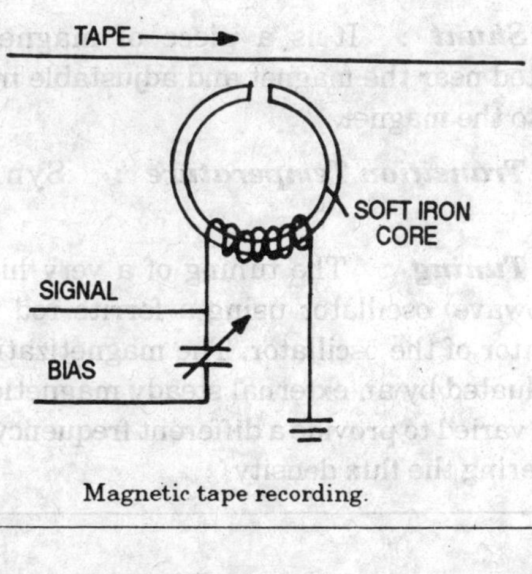

Magnetic tape recording.

distance between the pole pieces is the *gap length* and a good recorder may have a gap length as a small as 0.6 mm permitting a sharper record and thus more faithful reproduction. The magnetic tape completes the magnetic circuit.

Commercial tape recorders are generally either reel-to-reel or cassette types.

Magnetic recording is also employed for magnetic memories in computers and data-processing systems. Data can be stored on and retrieved from magnetic tape or a disk or drum.

Magnetic Screening : It is the application of a screen of high permeability magnetic material in order to protect electric circuits, devices, or other apparatus from the effects of magnetic fields. Magnetic screening is mostly employed to surround sensitive alternating-current measuring instruments and shield them from stray external magnetic fields.

Magnetic Shell : It means a thin sheet of ferromagnetic materials, like iron, magnetized across its thickness in such a way the two sides have opposite polarity. It may also be regarded as an infinite number of small bar magnets and the strength is expresseds as magnetic moment per unit area.

Magnetic Shift Register : It is a shift-register having an array of magnetic cores.

Magnetic Shunt : It is a piece of magnetic material mounted near the magnet and adjustable in position relative to the magnet.

Magnetic Transition Temperature : Syn. for Curie point.

Magnetic Tuning : The tuning of a very high frequency (microwave) oscillator using a ferrite rod in the cavity resonator of the oscillator. The magnetization of the rod is evaluated by an external steady magnetic flux density and is varied to provide a different frequency in the cavity by altering the flux density.

Magnetism : The phenomenona associated with regions having magnetic flux.

Magnetite : Syn. For lodestone. It is an ore of iron and exhibits ferrimagnetism.

Magnetization or Intensity of Magnetization : Symbol : M; unit: ampere per metre. A measure of the magnetic polarization which takes place when a material is placed in a magnetic flux. It is defined as the magnetic moment per unit volume and is the product of magnetic field strength, H, and magnetic susceptibility, X*m*:

M=HXm

M=B/μo-H

where B is the magnetic flux density and mo is the permeability of free space.

Magnetize: : To create a magnetic flux density in a material; to make a material to exhibit magnetic properties.

Magnetocaloric Effect of Thermomagnetic Effect : It refers to a fall in temperature occurring in certain paramagnetic salts when demagnetized adiabatically.

Magnetomotive Force : (m.m.f.) Symbol: Fm; unit: ampere; ampereturn. The line integral or magnetic flux density, *B,* for a closed path, divided by the magnetic constant mo which is represented by the equation:

Fm=mo-1fBdl

It is equal to the total conduction current linked, during one traverse of the closed path when the path encloses a current l the value of Fm increases by l for every traverse, *i.e. Fm* is multivalued unlike electromotive force, which is single-valued.

Magneto Resistance : It refers to a change in the resistance of a ferromagnetic material when it is subjected to a magnetic flux.

Magneto Striction : It is the mechanical deformation of a ferromagnetic material when subjected to a magnetic field.

Magnetostriction Oscillator : An oscillator which employs the phenomenon of magnetostriction to give oscillations or to control the frequency of an oscillator. In a simple magnetostriction oscillator a bar of mangetostrictive material is magnetized using a direct magnetic field and a coil or coils wound coaxially around it. A tuned circuit is connected in series with the coil. Direct voltage applied to the tuned circuit causes oscillations that are normally quickly damped. In case the frequency of the tuned circuit corresponds to the natural frequency of vibration of the magnetostrictive bar, the oscillations are maintained due to the interaction of the stresses in the bar and the associated changes of permeability.

A magnetostriction oscillator may be employed as a source of sound energy. The induced longitudinal vibrations of the bar are used to provide sound waves. These type of oscillators are very efficient when used under water.

Magnetostrictor or Magnetostriction Transducer : It is a transducer which utilizes magnetostriction to convert electrical energy into mechanical energy or vice versa. The general process is to clamp a rod or tube of ferromagnetic material at its centre and pass an alternating current through a coil wound around the length of the material.

Magnetron or Magnetron Oscillator : It refers to a crossed-field microwave tube that produces radiofrequency (r.f.) oscillations in the microwave region.

The common magnetron is made of a central cylindrical cathode surrounded by a cylindrical anode divided into two or more segments or containing several cavity resonators. A steady electrostatic field is applied between the anode and cathode. A steady magnetic flux density is applied parallel to the cylindrical axis and orthogonal to the electrostatic field. The magnetic flux is mostly provided by a permanent magnet or sometimes by an electromagnet. The symbol for the permanent magnet type is shown in Fig. a. Electrons, given out by the cathode, move under the influence of these two fields. The interaction of the electrons with gaps or the resonant cavities of the anode generates radio-frequency oscillations. These oscil-

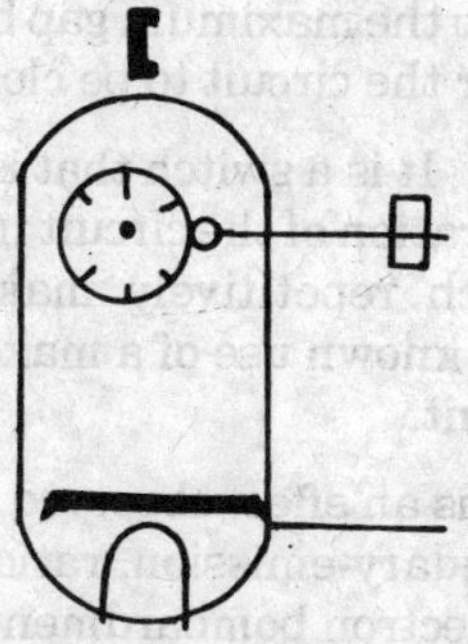

a. Symbol for magnetron with permanent magnet.

lations are given out through a coupled waveguide or coaxial line.

For a given value of anode voltage, the critical field is that value of magnetic flux density B at which an electron just fails to reach the anode and comes back to the cathode with zero kinetic energy. The *critical voltage* is the maximum anode voltage, in the presence of a fixed magnetic flux density, at which an electron just fails to reach the anode. The effect of an increasing magnetic field with fixed anode voltage is depicted in Fig. b.

Majority Carrier : It refers to the type of charge carrier in an

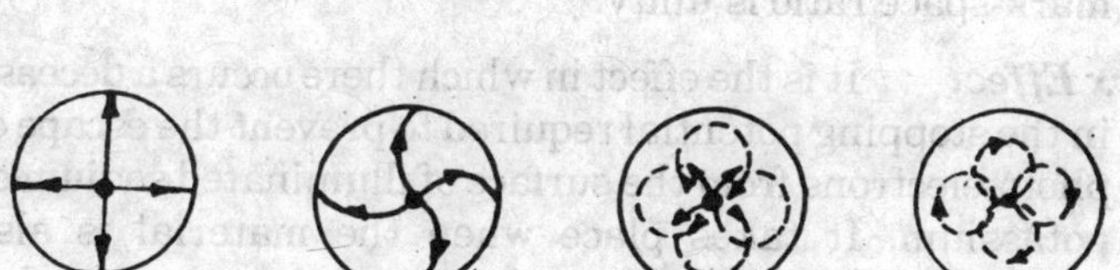

No. field Weak field Critical field Strong field

(b) Effect of magnetic field on electrons in magnetron.

Majority Carrier : It refers to the type of charge carrier in an extrinsic semi-conductor which constitutes more than half of the total charge carrier concentration.

Make : It refers to close a circuit by using a switch, circuit-breaker, or similar device.

It also refers to the maximum gap between the contacts of such device for the circuit to be closed.

Make-and-break : It is a switch that is automatically activated by the operation of the circuit in which it is incorporated, and which repetitively makes and breaks the circuit. The best known use of a make-and-break is in an electric bell circuit.

Malter Effect : It is an effect observed when a semiconductor or high secondary-emission, ratio, like caesium oxide is subjected to electron bombardment.

Manganese : Symbol: Mn. It is a metal having atomic number 25, used in alloys such as manganin and Heusler alloys. The element is also used in some forms of primary battery.

Manganin : It is an alloy containing 70.86% copper, 15.25% manganese, and 2.5% nickel and has a high coefficient of electrical resistivity, low temperature coefficient of resistance, and low contact potential.

Marconi Aerial : It refers to originally the simple vertical wire aerial, earthed at its lower end, used by Marconi. The term is now used for vertical aerials one quarter wavelength long.

Make-space Ratio : It is the ratio in a pulse waveform of the pulse duration to the time between successive pulses. It should be noted that in a perfect square wave the mark-space ratio is unity.

Marx Effect : It is the effect in which there occurs a decease in the stopping potential required to prevent the escape of photo-electrons from the surface of illuminated sodium or potassium. It takes place when the material is also illuminated by light of power frequency than that causing the photoelectric release.

Maser : Acronym from Microwave Amplification by stimulated emission of radiation. It is regarded as a source of intense coherent monochromatic radiation in the microwave region of the electromagnetic spectrum. The maser can be used as a microwave amplifier or oscillator.

There are two important types of solid-state maser-the

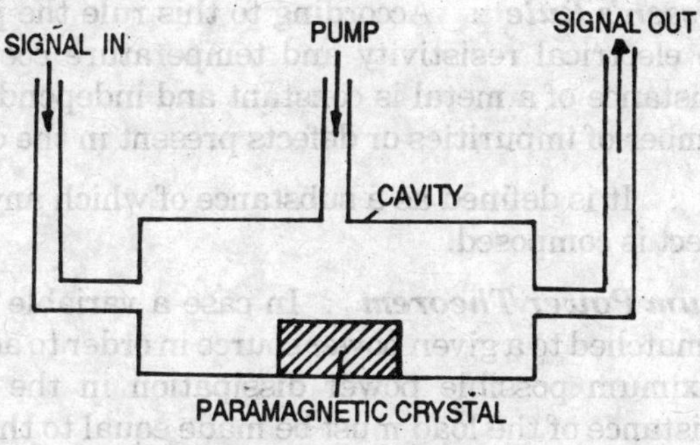

Sold-state cavity maser.

cavity maser and the travelling-wave maser. In the *cavity maser* material is placed in a cavity resonator and the pump and signal waves input to the cavity (*see* Fig). In the *travelling-wave maser* the maser material is placed in a waveguide which carries the pump power.

Microwave maser amplifiers are generally employed in radio and radar astronomy radiometer, terrestrial, communications, telemetry, satellite tracking and communications, and space communication.

Mask : It refers to a device used to shield selected ares of a semi-conductor chip during the manufacture of semiconductor components and integrated circuits.

It also refers to device made from metal foil and employed during the manufacture of thin-film circuits in order to define the pattern of material deposited as a thin film by vacuum evaporation on the substrate.

Mass Resistivity : The product of the mass and resistance of a conductor divided by the square of its length is known as the mass resistivity. It is equal to the product of resistivity and density of the material.

Master : See recording of sound.

Master Oscillator : It is an oscillator of extremely high inherent frequency stability which is employed to drive a power amplifier in order to supply a substantial power output to the load.

Matched Termination : It refers to a termination to a network or transmission line at which no reflected waves are obtained.

Mathiessen's Rule : According to this rule the product of the electrical resistivity and temperature coefficient of resistance of a metal is constant and independent of the number of impurities or defects present in the crystal.

Matter : It is defined as a substance of which any physical object is composed.

Maximum Power Theorem : In case a variable load is to be matched to a given power source in order to achieve the maximum possible power dissipation in the load, the resistance of the load must be made equal to the internal resistance of the source. The *available power* is given by is V°2/4Ri where V° is the open circuit electromotive force of the source and Ri the internal resistance.

Maxwell : Symbol: Mx. It is the unit of magnetic flux in the CGS electromagnetic system of units. One Maxwell equal 10-8 weber.

Maxwell Bridger : It is as four-arm bridge for measuring inductance in terms of a capacitance and resistance (*see* diagram). The balance point is indicated by a null response on the instrument, l,

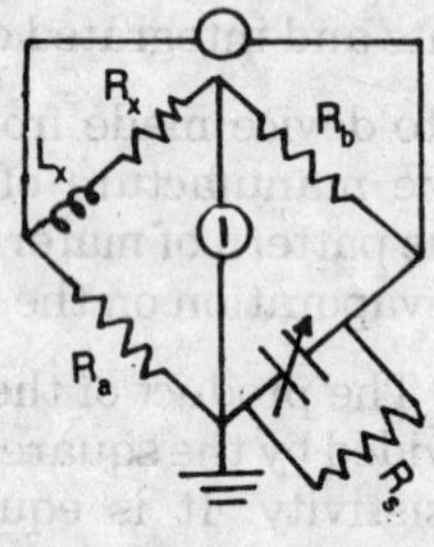

Maxwell bridge

$R_3R_x=R\delta R\propto$

$L\propto=R\delta R\propto C_3$

Maxwell's Equations : They refer to a set of classical equations relating the vector quantities applying at any point in a varying electric or magnetic field. The four basis equations are represented as:

Curl $H=\partial D/\partial t+j$

div B=O

curl E = $-\partial B/\partial dt$

div D=ρ

where h is the magnetic field strength, D the electric displacement, t is time, j is the current density, B the magnetic flux density, E the electric field strength, and r the volume charge density.

From these equations Maxwell derived that each field vector obeys a wave equation; he also proved that in free space, where j=O and r=O, the solutions represent a transverse wave travelling through space with the velocity of light.

Maxwell's Rule : According to this rule an electric linked circuit by a magnetic flux each part of the circuit experiences a force making it to tend to move in such a direction as to enclose the maximum possible magnetic flux.

Mean Life : It refers to the mean time to failure of a device, component, or any part or sub-system which can be separately tested. For any particular item the mean life will be based on the results of life tests and one or more of the following will be quoted.

The *observed mean life* is regarded as the mean value of the observed times to failure of all specimens in a sample of items under stated stress conditions. The criteria for what constitutes failure should be stated.

The *assessed mean life* is regarded as the mean life of an item determined as a limiting value of the confidence interval with a stated probability level based on the same data as the observed mean life of a nominally identical items. The following conditions apply: the source of the data should be stated; results may be combined in case when all conditions are similar; it should be stated whether one or two-sided intervals are being used; the lower limiting value is generally used for mean life statements and the assumed underlying distribution should be stated.

The *extrapolated mean life* is regarded as the extension of the observed or assessed mean life by a defined extrapolation or interpolation for stress conditions different from those applying to the conditions of the assessed mean life.

The elapsed time at which a stated proportion (*q* per cent) of a sample or population of items has failed is known as the *q-percentile life.* This is stated to be the observed percentile life, assessed percentile life, and extrapolated percentile life under similar conditions to those quoted for mean life.

Medium Voltage : A voltage having range 250 to 650 volts:

Medium Wave : A radiowave having wavelength in the range 0.1 to 1 Km, *i.e.*, of frequency between 3000 and 300 kilohertz.

Mega : Symbol: M. It means a prefix to a unit, denoting a multiple of 106 (*i.e.*,1 048576): one megabyte equals 2 20 bytes.

Megaphone : It is portable device employed to amplify and direct sound. It consists of a microphone, amplifier, and loudspeaker in a conical horn together with a battery power supply and acts as a directional loudspeaker.

Megger : It refers to a portable insulation tester calibrated directly in megohms.

Meltback Transistor : It refers to a type of bipolar junction transistor having the junctions formed by melting the semi-conductor, doping to the required polarity, and then allowing it to solidify.

Memory or Store : Any device associated with a computer which is employed to store information, such as programs or data, in digital form. The minimum region of any memory device required to store one unit of information is called a *storage element*.

Memory Location or Storage Location : A storage element which has a unique address in a memory.

Mercury : Symbol : Hg. It is a metal having atomic number 80, which is the only metal that is a liquid at room temperatures. It is used in electronics as the cathode in mercury-arc rectifiers, and as an amalgam in some electrolytic cells.

Mercury-arc Rectifier or Pool Rectifier : A cold-cathode arc-discharge tube which contains a mercury-pool cathode. When a sufficiently high voltage is applied to the anode an arc is struck across the tube. As the alternating-current input is applied the arc oscillates between anode 1 and anode 2 as each becomes positive in turn. Direct current is output to the load.

Mercury-Vapour Lamp : An arc-discharge tube which is used as a lamp. The gas in the tube is mercury vapour and the arc is struck between mercury electrodes. It should be noted that the light of the mercury arc is very rich in ultraviolet radiation and the tube can be used as flourescent lamp by coating the inside of the glass tube with a fluorescent powder, which converts the ultraviolet radiation to visible light.

Mercury-Vapour Rectifier : It refers to a gas-filled rectifier tube in which the cathode is in the form of a hot wire and the gas used is mercury vapour.

Mesa Transistor: : A bipolar junction transistor having the base which is diffused nonselectively into a suitable substrate and the portions around the base then etched away to leave a plateau above the substrate. A double diffusion technique may be employed to get the emitter inside the base region or the emitter may be alloyed into the base.

Mesfft : Acronym from metal semiconductor field-effect transistor. *Syn.* Schottky-gate field-effect transistor. It is regarded as a junction field effect transistor that has a Schottky barrier as the gate electrode rather than a semiconductor junction.

Metal Film Resistor : It refers to a film resistor which employs a metal as the resistive element.

Metallic Circuit : A circuit composed entirely of metallic conductors which connects two points in a telecommunication system and which does not employ an earth return.

Metallic Crystal : A crystal containing a regular array of positive ions together with a *cloud* of free electrons. Elements which involve a *metallic bond* of this nature are good electrical conductors.

Metallizing or Silvering : Depositing thin films of a metal (originally silver) on a glass, semiconductor, or other substrate in order to render it electrically conducting.

There are many methods employed for metallizing. The selected method depends on the substrate and the metal to be deposited. Chemical reduction is generally employed to deposit silver. Electroplating, vacuum evaporation and cathode sputtering are some other methods employed.

Metal Rectifier : A rectifier consisting of metal in contact with a suitable solid, like a semiconductor or an oxide of the metal.

Meter : Any instrument, such as a voltmeter or ammeter which is used for measuring.

Meter-Protection Circuit : A circuit used to protect a meter, like a voltmeter, from an overload current.

Meter Resistance : It refers to the internal resistance of a mater, measured at its terminals at a given temperature.

Metre : Symbol:m. The SI unit of length defined as the length equal to 1 50 763.73 wavelengths in a vacuum of the radiation associated with the transition between the levels 2p10 and 5d5 of the krypton-86 atom. The wavelength of this radiation is about 605.8 nanometres. The metre can therefore be measured very accurately and, unlike the original *international prototype metre,* which was a bar of platinum, does not change with time.

Metre Bridge : Wheatstone bridge having the ratio arms in the form of a tappable uniform wire of length one metre.

Mho : Symbol : Ω^{-1}. The reciprocal ohm, formerly used to measure conductance but now replaced by (and equal to) the siemens.

Mica : A naturally occurring mineral which consists of complex aluminium-potassium silicates and has a monoclinic structure; it can therefore by readily cleaved into thin plates. It is widely used for electrical insulation due to its unequalled combination of physical and electrical characteristics.

Mica Capacitor : A capacitor which employs mica as the dielectric material. Mica capacitors have characteristics of low loss, low temperature coefficient of capacitance, and good frequency stability.

Micro-Symbol: μ. : It is used as a prefix to a unit, and denotes a submultiple of 10^{-6} of that unit: one microsecond equals 10^{-6} seconds.

Microalloy Transistor : A type of bipolar junction transistor containing the semiconductor substrate which forms the base; the emitter and collector are formed by recrystallization from a suitable impurity-metal semiconductor alloy in small pits etched in the surface of the base. *Microalloy diffused transistors* are obtained in a similar manner but a gaseous diffusion is carried out first to produce a non-uniform base before the emitter and collector are formed. These transistors are longer used.

Microelectronics : it is the branch of electronics which deals with or applied to the realization of electronic circuits or systems from extremely small electronic parts. It studies the design, production, and application of any *microminiaturization* technique to reduce the cost, size and weight of electronic parts sub-assemblies, and assemblies and to replace vacuum-tube circuits with solid-state compatible parts.

Micron : Symbol: μ. It is a unit of length equal to 10^{-6} metre, now called the micrometre (mm).

Microphone : It is an instrument which converts sound energy into electrical energy. It forms the first element in a telephone, a broadcast transmitter, and all forms of electrical sound recording.

There are many different types of microphones; the most common types are the carbon, capacitor, crystal, moving-coil, and ribbon microphones. Many types of microphone operate by converting the sound waves into mechanical vibrations which then given electrical energy.

Microprocessor : It is regarded as the physical realization of the cental processing unit of a given computer system on either a single chip or semiconductor or on a small number of chips.

Microprogram : A program employed to define the instruction set of a computer.

Microstrip : It is a strip of conducting material in close proximity to an earthed conducting plane; it is employed as a transmission line in ultra-high frequency applications.

Microwave : It is an electromagnetic wave having wavelength in the range three millimetres to 1-3 metres, i.e., between infrared radiation and radiowaves on the frequency spectrum.

Microwave Generator : The generator like a klystron or magnetron, employed to generate microwaves.

Microwave Tube : It is an electron tube which is suitable for use as an amplifier or oscillator at microwave frequencies. These tubes mostly employ velocity modulation of the electron beam rather than density modulation as in the valves used in audiofrequency valve amplifiers or oscillators.

Microwave tubes are of two main types: *linear beam tubes, and crossed field tubes.*

Milking Generator : It is regarded as a low-voltage d.c. generator which is employed to charge one or more cells of an accumulator independently of the others.

Miller Effect : It refers to the phenomenon by which an effective feedback path between the input and output of an electronic device is provided by the interelectrode capacitance of the device. It means total input admittance of the device, can be affected which results in the total dynamic input capacitance of the device being always equal to or greater than the sum of the static electrode capacitances.

Miller Integrator : An integrator containing an active device, like a transistor, in order to improve the linearity of the output from a pulse generator. Miller integrators are generally employed with sawtooth pulse generators, like those used to generate a time base.

Milli Symbol : It is a prefix to a unit, denoting a submultiple of 10-3 of that unit: one millimeter equals 10-3 metre.

Minimum Discernible Signal (mds) or Threshold Signal : It is the smallest value of input signal to any circuit or device which just produces a discernible change in the output.

Minority Carrier : The type of charge carrier in an extrinsic semiconductor which constitutes less than half of the total charge carrier concentration.

Mismatch : It is the condition which takes place when the impedance of a load does not equal the output impedance of the source to which it is connected.

Mixer or Frequency-Changer : A device which is employed in superheterodyne reception to produce an output signal of different frequency from the input signal.

MKS System : A system of absolute units having the fundamental units of mass, length, and time are the kilogram, metre, and second and the fourth fundamental quality necessary to define completely electric and magnetic quantities is the permeability of free space, mo. The permeability was originally defined as having the value 10^{-7} henry/metre; the *rationalised MKS system*, is the basis of the system of SI units.

Mode of Transmission Mode : It refers to any one of the several different states of oscillation of an electromagnetic wave of given frequency. The mode of an electromagnetic wave depends upon the configurations of the vectors describing the wave; three main types of wave exists:

TE *waves (or h waves)* are transverse electric waves containing the electric field vector (E-vector) which is always perpendicular to the director of propagation, z. i.e.,

$$Ez=O$$

TM waves (or E *waves)* are transverse magnetic waves containing the magnetic vector (H-vector) which is always perpendicular to the direction of propagation, z. i.e.,

$$Hz=O$$

TEM waves are transverse electromagnetic waver containing both the E vector and the h vector which are perpendicular to the direction of propagation, i.e.

$$Ez=Hz=O.$$

The TEM mode is the one most commonly excited in coaxial lines. It cannot be propagated in a waveguide.

Modem : Acronym from modulator-demodulator. It is a device which is employed to convert signals output from one type of equipment into a form suitable for input to another type.

Modulation : The alteration or modification of any electronic parameter by another.

Modulator : 1 It refers to a device that effects modulation.

2 Or *Master Trigger*. It is generally a device like a multivibrator, which is used in radar to generate a train of short pulses, each of which acts as a trigger for the oscillator.

Modulator Electrode : A electrode which is employed to modulate the flow of current in an electrode device.

Monochromatic Radiation : The electromagnetic radiation of a single frequency is known as monochromatic radiation.

Monoscope : A type of electron tube which is employed to give single images, like a test pattern, suitable for television broad-casting. it contains an aluminium plate electrode with the desired image printed on it in carbon. The plate is scanned using an electron beam and video signals are produced due to the difference in secondary emission of electrons of aluminium and carbon.

Monostable or One-shot Univibrator : It refers to a type of circuit which contains only one stable state but on the application of a trigger pulse it can take up a second quasi-stable state.

Morse Code : It is regarded as an internationally agreed code for the transmission of signals, particularly in telegraphy.

A *Morse key* is a manually operated device which is employed to convert information to Morse code. A *Morse printer* is a device which converts received information, recorded on a suitable medium, into the corresponding characters.

Morse Telegraphy : Telegraphy containing the signals which are transmitted as a Morse code.

Mosaic Crystal : It refers to an imperfect crystal composed of a number of small crystals which are grown together so that the corresponding crystal planes are nearly or exactly parallel but have discontinuities as their mutual surfaces.

Mosaic Electrode : It refers to the light-sensitive electrode of a television camera tube on which the image is obtained.

MOS Capacitor : The basic metal-insulator-semiconductor structure which is made of a an upper *gate electrode* separated from the surface of a semiconductor substrate by a thin layer of insulating material (Fig a). The structure is the fundamental part of all MOS devices and integrated circuits like insulated-gate field effect transistors, charge-coupled devices, and MOS random-access memory. The behaviour of all MOS devices depends on the conditions present at the surface of the semiconductor substrate : the MOS capacitor is an extremely helpful tool for investing such surfaces as the total capacitance of the device depends on the surface conditions. devices depends on the conditions present at the surface of the semiconductor substrate : the MOS capacitor is an extremely helpful tool for investing such surfaces as the total capacitance of the device depends on the surface conditions.

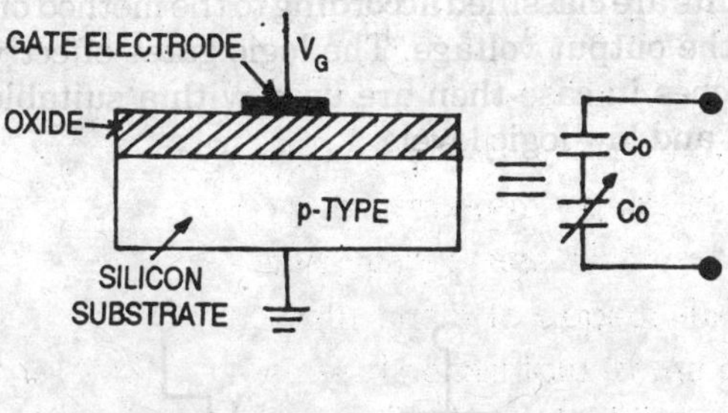

(a) MOS capacitor and equivalent circuit.

The total capacitance of the device is a function of the voltage Vc, applied to the gate electrode and is also frequency dependent. An applied potential difference V/G, across the device appears partly across the oxide layer and partly across the semiconductor as represented by the equation

$$Vg=Vo+fs$$

where Vf is the potential difference across the oxide and

f is the potential difference across the silicon. By convention the bulk potential of the silicon is zero and f/s is thus termed as the *surface potential.*

The capacitance Co of the oxide layer is constant and is represented as

C_o-eo_xd

where *eod* is relative permittivity of the oxide and *d* is the thickness. The capacitance CD of the silicon is a function of the surface potential fs. The variation of fs with applied voltage V_G is depicted in Fig.b.

MOS Integrated Circuit : An integrated circuit based on insulated gate field-effect transistors. MOS circuits have many advantages compared to bipolar integrated circuits for a substantial proportion of all semiconductor devices produces.

MOS Logic Circuit : It refers to a logic circuit constructed in an MOS integrated circuit. It consists of a combination of MOS field-effect transistors in series or in parallel that perform the logic functions. In other words it acts as AND or OR gates, etc. (Fig.a), coupled to other MOS transistors which give the output voltages of the circuit. MOS logic circuits are classified according to the method of determining the output voltage. The logic gates effectively act as switches in case then are used with a suitable choice of high and low logic levels.

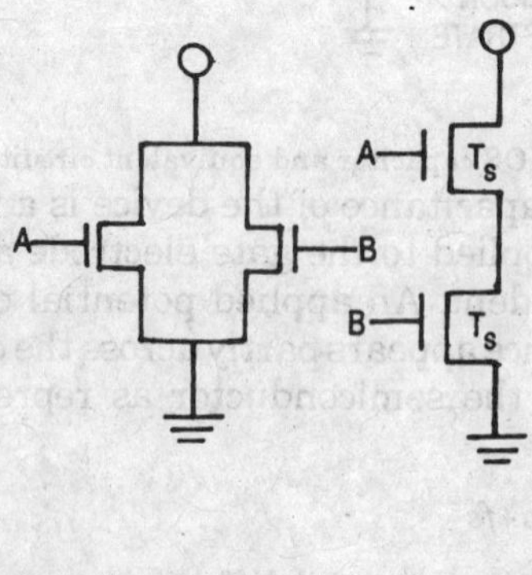

a MOS OR function AND function

In the case of a *ratio circuit* the logic gate, represented as a single switch transistor, T/L is connected in series with a load transistor T/L. The drain of the load transistor is connected to the power supply and the source of the switch transistor to earth (Fig.b).

In this type of static operation the voltage is applied continuously; in dynamic operation the voltage is applied on the application of a clock pulse in order to reduce dissipation.

MOS Transistor. (MOST;MOSEFET) : It means an insulated-gate field-effect transistor having the gate electrode which is metal (either aluminium or very high doped polysilicon), the insulator is silicon dioxide and the substrate is silicon.

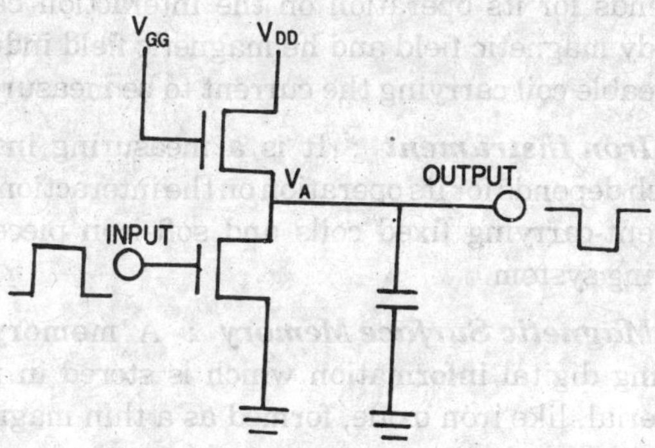

b Ratio circuit.

Motional Impedance or Driving Impedance : Driving-point impedance. A component of the impedance of an electromechanical or acoustic transducer which is due to the counter e.m.f. produced by the motion of the transducer. It should be noted that it is equal to the vector difference between the input impedance measured under specified load conditions and the blocked impedance.

Motor : A machine which converts electrical energy into mechanical motion.

An alternating-current (a.c.) motor is operated by using an alternating-current power supply. The rotating parts are termed the *armature (or rotor)* and the stationary windings the *stator*. A direct-current (d.c) motor is operated by using a direct-current power supply and usually contains a *commutator* which connects each of the sections of the primary winding in turn to the power supply in order to provide the necessary torque.

Motorboating : Noise which occurs in low-frequency and audio-frequency amplifiers and is caused by undesired oscillations arising in the amplifiers. It results in a sound from the loudspeaker which resembles a motorboat engine.

Moving-Coil Galvanometer : It is based on the principle that a conductor carrying current when placed in a magnetic field experience a force in the direction given by Fleming left hand rule.

Moving-Coil Instrument : A measuring instrument which depends for its operation on the interaction between a steady magnetic field and he magnetic field induced in a moveable coil carrying the current to be measured.

Moving-Iron Instrument : It is a measuring instrument which depends for its operation on the interaction between current-carrying fixed coils and soft-iron pieces in the moving system.

Moving Magnetic Surface Memory : A memory device having digital information which is stored in magnetic material, like iron oxide, formed as a thin magnetic film on a nonmagnetic substrate.

Moving-Magnet Instrument : It refers to a measuring instrument which depends for its operation on the interaction between the magnetic flux density induced in a fixed coil by the current to be measured and a small permanent magnet suspended in the plane of the coil.

Multichannel Analyser : It refers to a thermionic value which contains more than three electrodes.

Multielectrode Valve : It refers to a thermionic value which contains more than three electrodes

Or Multiple-Unit Valve. It also refers to thermionic valve which contains within one envelope two or more sets of electrodes each having its own independent stream of electrons.

Multiplex operation : It is regarded as the operation in which like simultaneous transmission of several signals takes place along a single path without any loss of identity of an individual signal.

Multivibrator : An oscillator having two linear inverters coupled in such a way that the output of one provides the input for the other. There are large types of multivibrator, the action of which depends on the type of coupling used.

Capacitive coupling produces an *astable multivibrator* which contains two quasi-stable states; once the oscillations are established the device is free-running.

Resistive capacitive coupling produces a monostable multi-vibrator.

Resistive coupling (see diagram) generates a bistable circuit which has two stable states and can change state on the application of a trigger pulse.

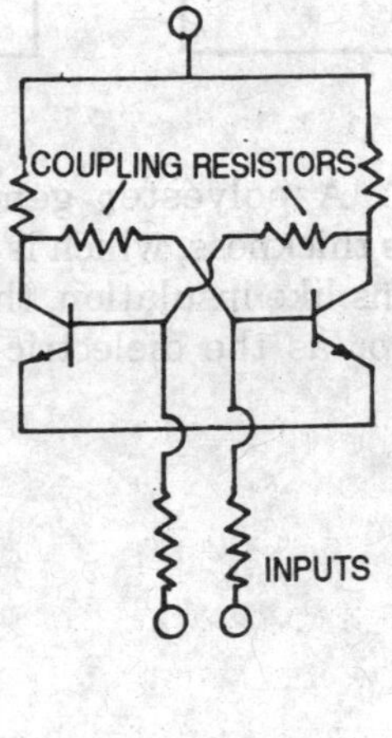

Bistable multivibrator

Muting Switch : It refers to a switch which automatically activates a noise-suppression circuit in a particular piece of equipment when required, as when the noise in the system exceeds a predetermined level.

Mutual Capacitance : It is an indication of the extent to which two capacitors can interact.

Mutual Conductance or Transconductance
: Symbol:$g/_m$. The ratio of the incremental change in the output current *Iout*,of any amplifying circuit or device, to the increment change of input voltage, Vin, causing it, when the output voltage, Vout, is held constant.

Mutual Inductance : Consider two coils A and B adjacent to each. When the current in the circuit A changes; there is change in the magnetic flux linked with it and an emf is induced in the circuit B. This phenomenon is called *mutual induction,* and the pair of circuit which show it are said to have *mutual inductance.*

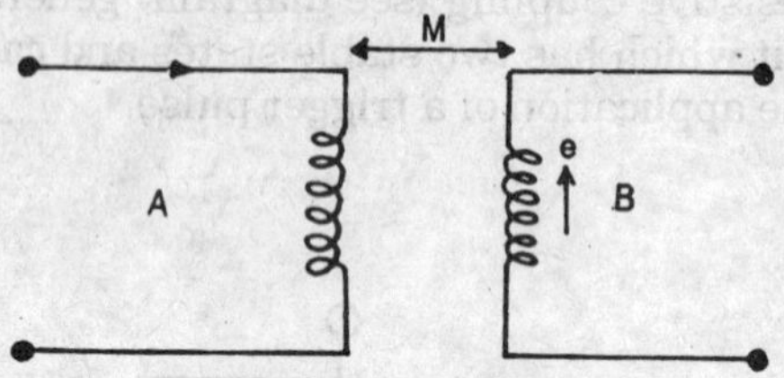

Mylar, *Tradename* : A polyester generally obtained in sheets of various thickness, which is employed for a variety of applications like insulation, the base material of a magnetic tape, or as the dielectric in certain types of capacitor.

N

Nano : It is a prefix to a unit, denoting a submultiple of 10^{-9} of that unit: one nanometre equals 10^{-9} metre.

NASA : It is *abbrev.* for National Aeronautical and Space Administration.

Nature Frequency : It is defined as the frequency at which free oscillations in an electrical or mechanical system.

N-Channel : It denotes an MOS circuit or device, such as a charge-coupled device, MOS integrated circuit, or MOS transistor, in which the conducting channel is formed as *n*-type semiconductor.

Negative Bias : It is a voltage that is applied to an electrode of an electronic device and is negative with respect to some fixed reference potential, usually earth potential.

Negative Resistance : It is the property certain devices of whereby a portion of the current-voltage characteristic has a negative slop, *i.e.*, the current decreases with increasing applied voltage.

Neon : Its symbol is Ne. It is an inert gas, atomic number 10, that exhibits a characteristic red glow when ionized and is extensively used as the gas in glow lamps. Neon lamps may be made extremely large for use in illuminated signs or as indicators and voltage regulators.

Neper : Symbol: Np. It is dimensionless unit used in telecommunications to express the ratio of two powers. It is the natural logarithm of the square root of the power ratio.

Nernst Effect : It is the effect of a conductor or semiconductor placed in a magnetic flux density and a temperature gradient maintained across it at right angles to the magnetic flux density. In other words an e.m.f. is developed in the material orthogonal to both the magnetic flux density and the temperature gradient.

Nerve Current : It is regarded as a naturally occurring extremely small current which flows through a nerve path in a human or other animal.

Net Loss : It is the difference between the attenuation and the gain in any circuit, device, network, or transmission line.

Network : It is a system in which a number of impedances connected together to form a system which has a set of inter related circuits and that performs specific functions.

Bilateral networks conducted in both directions whereas which conduct in only one direction are *unilateral.*

Passive networks has no energy source of sink other than normal ohmic losses; those which do contain an energy source or sink are *active.*

All-pass networks attenuate all frequencies equally; other networks are described according to their frequency response.

A point within a network at which three or more of the elements are joined is called a *branch point (or node)*;

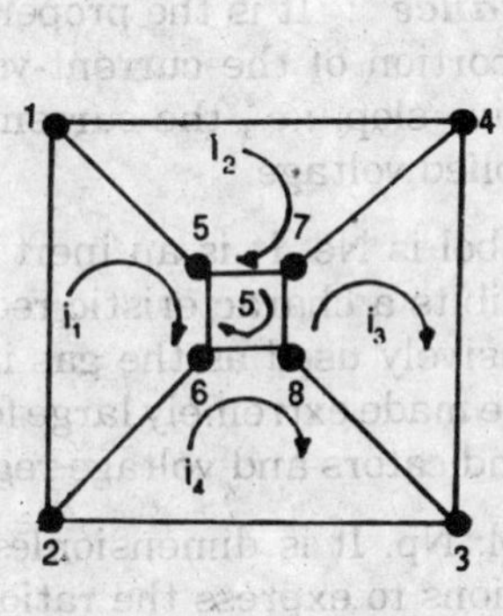

Conducting paths of a network.

points 1-8 in Fig.below) are branch points. A conducting path between two these type points is called a *branch* (1.2, 3, 4, etc. in Fig.). A voltage at a point in the network relative to the voltage at a designated branch point is termed a *nude voltage.* A closed conducting loop in the network (*e.g.* 1,3,5,7,1) is regarded as a *mesh contour,* and the portion of the network bounded by it is called a *mesh.* Any branch which has common two or more meshes is a *mutual branch (e.g.,5-6).* Two branches of a network are regarded as *conjugate* if an e.m.f. in one of them does not produce a current in the other.

The behaviour of a network may be analysed with the help of Kirchhoff's laws to each mesh in the network. in case of large networks with many meshes, as with many filter networks this is complicated.

Analysis of linear networks is carried out by considering the network as a quadripole and deriving sets of equations relation the currents, voltage, and impedances at the input and output.

Three different sets of equations can be written down; the impedance equations, the admittance equations, and derived from these the hybrid equations. The impence equations are given as below:

$$\begin{vmatrix} V_1 \\ \\ V_2 \end{vmatrix} = \begin{vmatrix} Z_{11} & Z_{12} \\ \\ Z_{21} & Z_{22} \end{vmatrix} \begin{vmatrix} i_1 \\ \\ i_2 \end{vmatrix}$$

It is called the Z-matrix. Equivalent matrices can be written for the hybrid and admittance equations. These are the H- and y- matrices, respectively. The constants in these equations are called the *matrix parameters.* Three terminal devices, like transistors, can be represented as quadripoles that have two terminals joined together.

Neuroelectricity : Electricity produced in the nervous systems of humans and other animals.

Neutral : It is having no net positive or negative electric charge; at earth potential

It also denotes the line that completes the domestic mains supply and is connected to earth at the lower station.

Neutralization : It is defined as the provision in an amplifier of negative feedback of that amplitude and phase which it counteracts any inherent positive feedback. If the positive feedback is not neuralized, undesired oscillations may occur.

The voltage feedback to the input is termed the *neutralizing voltage.* The degree of neutralization in an amplifier may be observed using an indicating device termed a *neutralizing indicator.*

Neutron : It is an elementary particle having the same mass as the proton, but it is electrically neutral which means it has no charge. Neutrons are present with protons in atom only in the nucleus, where they are stable. Free neutron are unstable. The absence of electric charge makes them difficult to detect but does not allow them to penetrate the open structure of atom.

Newton : Symbol : N. It is the SI unit of force defined as the force which when applied to a mass of one kilogram, gives it can acceleration of one metre per second.

Nichrome : It is an alloy containing approximately 62% nickel, 15% chromium, and 23% iron. Its resistivity is very high and can operate at extremely high temperatures. It is therefore suitable for use in making thin film resistors, wire-wound resistors and heating elements.

Nickel : Symbol: Ni. It is a metal, atomic number 28, and is widely used in electronics. It shows strong ferromagnetism and is used in magnetic alloys. It is also used as a conductor and in electrolytic cells.

Nickel-Cadmium Cell : It is defined as a secondary cell that has a nickel hydroxide/nickel oxide mixture as the anode, a cadmium cathode, and potassium hydroxide as the electrolyte.

Noctovision : It is regarded as a television system which views an infrared image rather than a visible light image and is sometimes used at night. A special designed infra-red-sensitive camera tube gives a video signal which may be detected by a standard receiver.

Node : It is defined as a point, line, or surface in a distributed field at which some specified variable of a standing wave, like voltage or current, attains a minimum value, usually zero. A *partial node* has a nonzero minimum. A point at which maximum magnitude is attained is an *antinode*.

Nodon Rectifier : It is defined as electrolytic rectifier containing an aluminium cathode, a lead anode, and aluminium phosphate solution as the electrolyte.

Noise : It is defined as any collection of undesirable and unwanted signals in an amplifier and other signal transmission there are various types of noise such as hum, cross talk, interference man-made noise and current noise among others.

In general, noise may be classified into two main types: *white noise* which contain a relatively wide flat frequency, spectrum, and *impulse noise,* which is due to a single momentary disturbance or a series of such disturbances.

Thermal noise is a form of white noise which is as a result of the random thermal, agitation of electrons in resistive components. *Random noise* which is due to any randomly occurring transient disturbance. In case rate of occurrence is sufficiently high it results in white noise, similar to thermal noise. In case rate of occurrence is low, random noise contributes to the impulse noise. It should be noted that all electronic circuits and device suffer from thermal noise and random noise. Additional sources of noise arise in telecommunications from many different sources. *Atmospheric* and *galactic noise* both contribute to radio noise. *Carrier noise* is due to undesired variations in the frequency or amplitude of a carrier wave that are not a result of the modulating wave.

Non-inductive : It denotes an electric circuit, that, for its desired function, has negligible inductance. A circuit which has zero inductance is extremely difficult to realize.

Non-reactive : It denotes any electric circuit, device, or winding which for its desired function, has negligible reactance.

Non-reactive Load : It is defined as a load in which the alternating current is in phase with the alternating voltage at the terminals.

Norton's Theorom : The theorem states that any network or circuit may be replaced by a constant current generator whose current output is equal to the network short circuit load current in shunt with an impedance equal to that seen looking back from the output terminals of the *network* with the original generates e.m.f. shorted.

N-type Conductivity : It is the conduction in a semiconductor in which current flow is due to the movement of electrons through the semiconductor.

N-type Semiconductor : it is a type of extrinsic semiconductor which has a higher density of conduction in electrons than of mobile holes, *i.e.* electrons are the majority carriers.

Nucleus : It is defined as the central and most massive part of an atom. It carries a positive charge Ze, where Z is the atomic number of the atom and *e* the electronic charge. A nucleus consists of tightly bound protons and neutrons collectively known as *nucleons;* the total number nucleons is called the mass number, A. The number of neutrons associated with a given number of protons can vary within limits, giving rise to various isotopes of an element.

Mostly naturally occurring isotopes have stable nuclei although some, together with all artificially produced isotopes, have unstable nuclei; these are radioactive and undergo chemical or nuclear transformation on decaying.

Nuclide : An atom which is completely characterized by its atomic number, Z, mass number, A, and nuclear energy state.

Null Method or balanced method : An accurate method of measurement in which the quantity being measured, like resistance or capacitance, is balanced by another of a similar kind: voltages in different circuits are adjusted in such a way that the response of an indicating instrument falls to zero. The best known example of the null method is in the Wheatstone bridge although it is employed in many other bridge circuits.

Null-Point Detector : An instrument which is employed with bridge circuits to give a zero response when the balance point is achieved. Direct-current bridge circuits employ a sensitive galvanometer: balance is achieved when a zero indication is found. Alternating-current bridge circuits frequently employ a sensitive loudspeaker, generally as a headset: the sound becomes inaudible at the balance point.

Number of Poles : These are defined as the number of different conducting paths that may be simultaneously opened or closed by a switch, circuit-breaker, or other similar apparatus. The device is regarded as single pole, double pole, triple pole, or multi pole depending on the number of poles that is can operate.

Nyquist Diagram : This is a diagram which can be employed as a criterion of stability in an amplifier. It shows the relation, in rectangular coordinates, between the amplification and the feedback of the device.

O

Octode : It is a thermionic valve which has a total of eight electrodes arranged as five grid electrodes between the cathode and the main anode and an additional anode placed between the two innermost grids. it is employed as a combined mixer and local oscillator.

Off-line : The term denotes peripheral device, possibly an on-line device, that is switched off, broken or disconnected from a computer.

Ohm : Symbol:Ω. The SI unit of electrical resistance is ohm. It is defined as the amount of resistance between two points on a conductor when a constant current of one ampere flows as a result of an external potential difference of one volt applied between the points. This unit was formerly called the *absolute ohm* and replaced the *international ohm* (Wint) as the standard or resistance: one Wint equals 1.000-9W.

It is also defined in term of energy, if one ampere of current passed through a resistance of one ohm for one second, one joule of energy will appear across the resistances in the form of heat.

Ohmic loss : Power dissipation in an electrical circuit, network, or device which is as a result of the resistance present rather than to other cause such as eddy currents or back e.m.f.

Ohmmeter : It is instrument that measures the electrical resistance of conductors or insulators. The indicating scale is calibrated in ohms. The meter is connected across in parallel with the device to be measured and a pointer indicates the resistance directly.

Ohm meter : Symbol Ωm. This is the SI unit of electric resistivity.

Ohm's Law : It states that the electric current, I flowing in a conductor or resistor is linearly proportional to the applied potential difference, V, across it. From the definition of resistance, R, Ohm's law can be written:

$$V = IR$$

Oil-break : It is a term denoting having contacts that separate in oil.

Omnidirectional Aerial : It is defined as an aerial whose radiative or receiving properties at any instant are the same on all bearings.

Ondograph : This is an instrument that produces a graph of an alternating voltage. it has a digital response and therefore the graph of the waveform appears as a series of small steps rather than as a smooth curve.

Ondoscope : It is regarded as a glow-discharge tube which detects the presence of high frequency radiation.

On Line : It denotes a device of a computer which is directly controlled by the computer and executes instructions as they are generated.

Open Arc : It is defined as an arc struck between carbon electrodes with the air being permitted free access to it. The products of the combustion tend to dissipate throughout the atmosphere rather than remaining in the immediate vicinity of the arc.

Open-Circuit Voltage : This is the voltage across the output terminals of any electrical or electronic network, device, machine, or other apparatus at no-load when designated operating conditions are carried out.

Open-ended : It denotes any circuit, system or process that can be readily augmented by the addition of further stages, parts, or steps.

Open Wire : it is *syn.* to overhead line. It is defined as a conductor that is not surrounded by insulation and is supported separately above ground.

Operating Point : It is the point on the family of characteristic curves of an active electronic device, like a transistor, that represents the magnitudes of voltage and current when designated operating conditions are applied to the device.

Operational Amplifier : The term is generally reserved for an amplifier of two or more stages that is designed for insertion into other equipment.

Opposition : Two periodic quantities of the same frequency are regarded as in *opposition (or antiphase)* in case the phase difference between them is one half-period *i.e.*, 180°.

Optical Ammeter : This is an instrument which measures the current flowing through the filament of an incandescent lamp by comparing, photometrically, the illumination produced with that produced by a current of known magnitude in the same filament.

Optical Character Reader : (OCR) It is a device which is used to produce coded signals suitable for a given digital

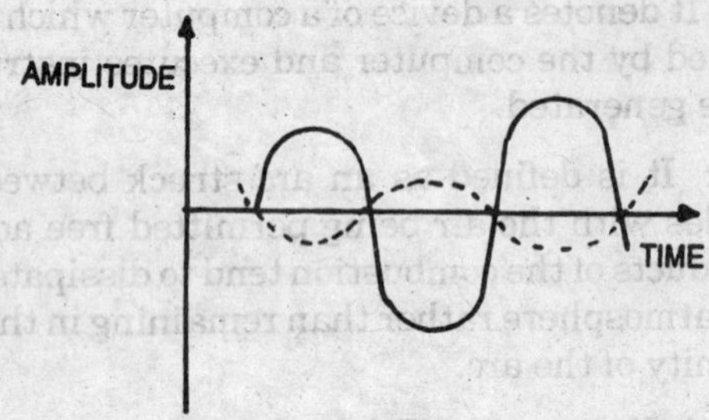

Opposition

computer from information in the form of characters, numbers, or other symbols printed on paper.

Optically Erasable memory : It is *syn. for* floating-gate PROM.

Oscillating Current (or Voltage) : A current (or voltage) waveform which periodically increases and decreases in amplitude with respect to time, according to a particular mathematical function. Oscillating waveforms can be sinusoidal, sawtooth, square, or many other shapes.

Oscillation : It is defined as a period variation of an electrical quantity, such as current or voltage.

It is also defined as a process which takes place in an electrical circuit if the values of self-inductance and capacitance in the circuit are such that an oscillating current results from a disturbance of the electrical equilibrium of the circuit.

A circuit which produces osilliations freely is called an *oscillatory circuit.* Oscillations which are due to the application of a direct voltage input to the circuit and continue until the direct voltage is removed are called *self-sustaining oscillations.* Oscillations which tend to decrease in amplitude with respect to time are called *stable oscillations; unstable oscillations* tend to increase in amplitude with respect to time and soon exceed to rated operating conditions of the circuit.

Oscillator : It is defined as a circuit which converts direct-current power into alternating-current power at a frequency which is generally greater than can be achieved by rotating electro-mechanical alternating-current generators. Application of the direct-voltage supply to the circuit is mostly sufficient to cause it to oscillate and for the oscillations to be maintained until the direct voltage is switched off.

There are two types of oscillators: *harmonic oscillators* generate essentially sinusoidal waveforms and contain one or more active circuit elements continuously supply power to the passive components.

The effective negative resistance is given out either by a device, such as a unijunction transistor, which exhibits a negative-resistance portion of its characteristic or by employing positive feedback of power in order to overcome the damping. Any particular oscillator may be studied from the negative-resistance approach or from a feedback approach. In the latter case internal positive feedback is said to be present in the negative resistance device. Generally *negative-resistance oscillators* are those which has a device like a unijunction transistor or tunnel diode (Fig. *a*), operated in the negative-resistance portion of the characteristic determined by the applied voltage, VA, and external source resistance, *Rs*. *Feedback oscillators* employ external positive feedback. An inherent phase shift of 180 takes place between the base and collector of the common-emitter connection.

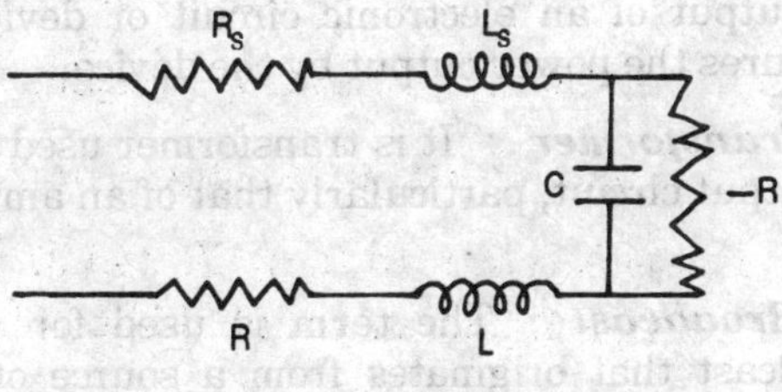

Equivalent circuit

(a) Negative-resistance oscillator.

Oscilloscope : It is an instrument which gives a visible image of one or more rapidly varying electrical quantities.

Ostwald's Dilution Law : This law gives the degree of dissociation of an electroyte in case it is dissolved in water. It is obtained by applying the law of mass action to electrolytic dissociation. Where unit amount of an acid, HA, is dissolved in a volume, V, of water it dissociates according to

$$HA \Leftrightarrow H^+A^-$$

At equilibrium, if the fraction of acid dissociated is ò, then the concentrations of acid and dissociated ions are given by the relation

$$\{H^+\}\{A^-\}/\{HA\}=\alpha^2/(1A)V=K$$

where K is a constant called the *dissociation constant* of the reaction. For weak electrolytes ò is very much less than unity and the law become to $\propto^2 = KV$. It means if the dilution of the electrolyte is increase the degree of dissociation also increases.

Outgassing : it is defined as the removal by heating of some of the air absorbed on the interior surfaces of a vacuum system or tube.

Output : It is the power, voltage, or current delivered by any circuit, device, or apparatus.

The term is also used for the part of any circuit, device, or apparatus, generally in the form of terminals, at which the power, voltage, or current is delivered.

Output Impedance : It is the impedance presented at the output of an electronic circuit or device.

Output Meter : This is an instrument which is coupled to the output of an electronic circuit or device and that measures the power output by the device.

Output Transformer : It is transformer used for coupling an output circuit, particularly that of an amplifier, to be load.

Outside Broadcast : The term is used for a television broadcast that originates from a source other than a television studio. An outside broadcast employs mobile cameras and transmitting or recording apparatus. A live outside broadcast employs a portable transmitter to replay the broadcast signal to the main control centre where it is broadcast from the main transmitters.

Overcurrent Release : It is a switch-circuit breaker, or other tripping device that operates when the current in a circuit exceeds a predetermined value. A current that causes the release to operate is an *overcurrent*

The device is mostly designed in such a way that delay takes place after the overcurrent is sensed and before the device trips.

Overdriven Amplifier : This is an amplifier that is operated with an input voltage greater than that for which the circuit was designed.

Overload : It is defined as any load delivered at the output of an electrical device, circuit, machine, or other apparatus which exceeds the rated output of the equipment.

Overvoltage : A voltage which exceeds the normal voltage applied between two conductors or between a conductor and earth, is called a voltage.

It is also the defined as the amount by which the applied e.m.f. necessary to release hydrogen on that electrode from a particular electrolyte exceeds the emf. required to liberate hydrogen from electrolyte on a standard platinum electrode.

Owen Bridge : It is a four-arm bridge used for the measurement of inductance in terms of known resistance and

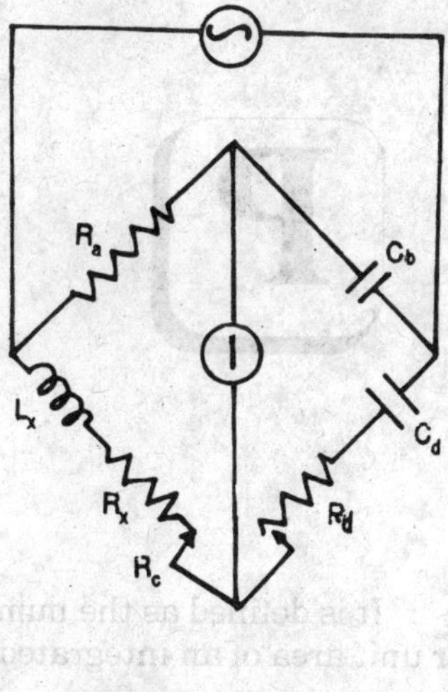

Owen Bridge

capacitance (see diagram). At balance by the null response of the indicating instrument, *I*,

$$L_x = C_o R_a R_a$$

$$R_x = (C_\beta R_a / C_d) - R_c$$

Oxidation : It is a chemical reaction employed in electronics in which a thin portion of the surface of a silicon ship is converted to silicon dioxide.

Oxide : It is a chemical compound of an element, with oxygen. In particular it is *short for* silicon dioxide (silica) It is the most widely used insulating and/or passivating material in the construction of devices, components, and integrated circuits fabricated in silicon. It should be noted that silicon dioxide can be readily grown on the surface of silicon and layers of varying thicknesses are grown during the fabrication of silicon devices and circuits.

Oxides are used as a barrier to the diffusion of impurities into the substrate material and is used as an 'on-chip' mask during the planar process.

Oxygen Overvoltage : (at an anode of an electrolytic cell). It is defined as the difference between the actual e.m.f. necessary to release oxygen at the anode and the theoretical volume for release of the gas.

P

Packing Density : It is defined as the number of devices or logic gates per unit area of an integrated circuit.

It is also the amount of information which is contained

in a given dimension of a storage system of a digital computer.

Pair : it is defined as transmission line consisting of two similar conductors which are insulated from each other but are associated to form part of a communication channel or channels. In case the two conductors are twisted around each other they form a *twisted pair*. A *coaxial pair* is a pair of cylindrical conductors which are coaxial.

Paired Cable : This type of cable is composed of several pairs of conductors: each pair is twisted together but no two sets of pair ate twisted.

Pairing : It is a fault present in television picture tubes which use interlaced scanning. Lines of alternate fields tend to coincide instead of interlacing with each other and the vertical resolution is halved.

PAL : It is abbrev. for phase alternation line. A colour-television system developed in Germany. The chrominance signal is resolved into two components in quadrature which are employed for amplitude modulation of the chrominance subcarriers. In the PAL system the relative phase of the quadrature components is reversed on alternate lines in order to minimize phase errors. This system is mostly used in Europe.

Pancake Coil : It is defined as a flat coil in which the common method of winding is to have the turns arranged as a flat spiral.

Panoramic Receiver : This is a radio receiver which is automatically tuned in such a manner to receive certain frequency bands, each for a preselected frequency. The period of the tuning variation is preselected.

Paper Capacitor : It is regarded as a capacitor of medium loss and medium capacitance-stability which is employed in high-voltage a.c. and d.c. applications. These capacitors are manufactured by winding together aluminium foils interleaved with layers of tissue paper. The moisture of the paper is removed by treating with a suitable oil or wax.

Paper Tape : It is strip of paper on which information can be carried in the form of holes punched across the width.

Parabolic Reflector : It is as a radiofrequency or microwave-frequency reflector which contains a hollow concave paraboloid shape in such a way that all waves passing through the focus will be reflected parallel to the axis of rotation.

Parallel : The circuit elements are said to be *in parallel* in case they are connected in such a way that the current divides between them and later reunites.

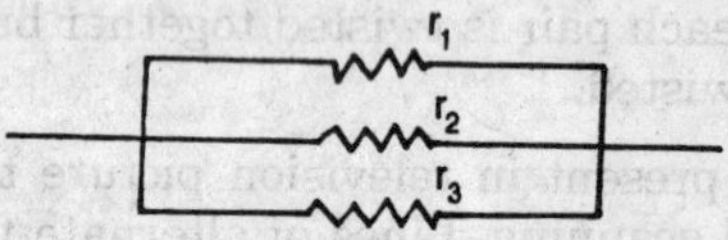

Resistor in parallels

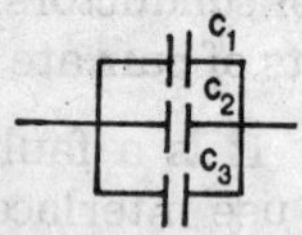

Capacitors in parallels

In case of *n* resistors in parallel the total resistance, *R* is given by the expression

$$IR = Ir1 + Ir2 + \quad 1rn$$

where $r_1, r_2 \ldots rn$ are the values of the individual resistors.

In case of n capacitors in parallel the total capacitance, *C*, is given by the expression

$$C = c_1 + c_2 + c_3 + \ldots cn$$

where $c_1, c_2 \ldots c_n$ are the individual capacitances. In other words, the capacitors thus behave collectively as a large capacitor having the total plate area of the component capacitors.

Machines, transformers, and cells are regarded to the in parallel when terminals of the same polarity are connected together. Several cells connected together in par-

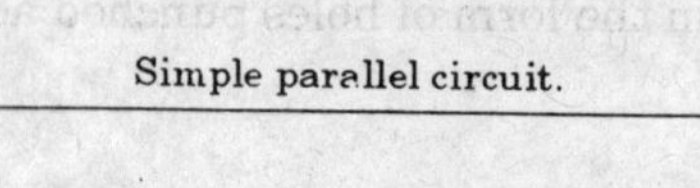

Simple parallel circuit.

allel have a lower total internal resistance than single cell and can thus supply a larger maximum current.

Parallel Circuit : It is a circuit having two or more elements connected in parallel across a pair of lines or terminals.

Parallel Transmission : It is defined as a communication system in which the characters of a word are transmitted in parallel along separate lines. Transmission of the characters is generally simultaneous.

Paramagnetism : This is regarded as an effect observed in certain materials having permanent atomic or molecular magnetic moment. Each orbital electron in an atom constitute an individual current and therefore has a magnetic moment; however the possible energy levels available are such that only unfilled shells of electron contribute to the magnetic moments of the atom as a whole. It is seen that free atoms have a magnetic moment due to orbital electrons in unfilled outer shells but practical substances combine in general so as to complete the outer shells; most gaseous molecules and ionic or homopolar liquids or solids have no overall magnetic moment and are diamagnetic. Permanent magnetic moments are only possessed by molecules or ions having unpaired electron spins or by particular ions of multivalent transition elements in which there is an unfilled inner shell. Paramagnetism is generally associated with electron spin but a few compounds also have an orbital contribution.

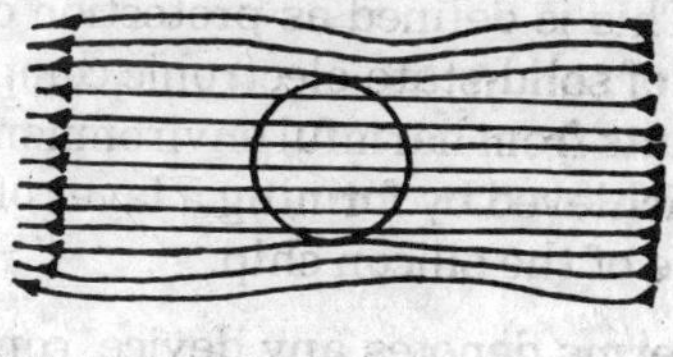

Change in magnetic flux density in a paramagnetic substance.

It is found that paramagnetism results in an increase of magnetic flux density within a sample; this is represented schematically by a concentration of the lines of magnetic flux density passing through it (Fig). When a paramagnetic substance is placed in a nonuniform magnetic field it tends to move from the weaker to the stronger region of the field; a bar of paramagnetic material placed in a uniform magnetic flux tends to orientate itself with the longer axis parallel to the flux.

Parameter : It is defined as a quantity that is fixed in a given case but has a particular value for each different case considered.

Parametric Amplifier : It is defined as microwave amplifier in which the reactance of the device is varied with respect to time in a regular manner. An alternating voltage, the pump voltage, is the most common from of *pump*.

Parasitic Oscillations : The undesired oscillations that may occur in amplifier or oscillator circuits. The frequency of such oscillations is generally very much higher than the frequencies for which the Circuit has been designed as it is mainly determined by stray inductances and capacitances, like in connecting leads, and by inter electrode capacitances.

Partition Noise : It is regarded as the noise that occurs in active electronic devices due to random fluctuations in the distribution of the currents between the electrodes.

Pascal : Its symbol is P_a. This is the SI unit of pressure, defined as the pressure which results when a force of one newton acts uniformly over an area of one square meter.

Paschen's Law : It states that the breakdown voltage which initiates a discharge between electrodes in a gas is a function of the products of pressure and distance.

Passivation : This is defined as protection of the junctions and surfaces of solid-state electronic components and integrated circuits from harmful environments. Passivation is generally achieved by forming a layer of silicon dioxide on the surface of the silicon chip.

Passive : The terms denotes any device, component, or circuit which does not introduce gain or does not have a directional function. In practice only pure resistance, capacitance, inductance, or a combination of these three is passive.

Passive Substrate : It is a substrate, like glass or ceramic, which is employed in microelectronics for its lack of transistance.

Patching : This is the forming of temporary connections between circuits of an analog computer in order to simulate a particular psychical process.

Suitable patching of chosen circuits in an analog computer creates in the establishment of a desired relationship between the voltages in the computer.

Pauli Exclusion Principle : It states that no two electrons in an atom can exist in the same quantum energy state, in other words they cannot be described by the same four quantum numbers, *l,m,n,* and *s*. A particular atomic orbital of given values of , *I*, and *m* can contain only two electrons : these electrons must have spins equal to $+\frac{1}{2}$ and $\frac{1}{2}$. An atomic orbital which has tow electrons of opposing spins is said to contain spin-paired, or paired, electrons. An orbital containing only one electron is said to contain an unpaired electron, or unpaired spin.

P Band : It is regarded as a band of microwave frequencies ranging from 0.225 to 0.390 gigahertz.

PBX : It is *abbrev.* for private branch exchange.

P-Channel : It denotes an MOS circuit or device, such as a charge-coupled device. MOS integrated circuit, or MOS transistor, in which the conducting channel is formed as *p*-type semiconductor. *N*-channel devices are however preferred. The term also describes junction field-effect transistors which have a *p*-type channel.

PCM : It is *abbrev.* for pulse code modulation.

P.D. : (or pd) : It is *abbrev.* for potential difference.

Peak Factor : It is defined as the ratio of the peak value of a periodically varying quantity to the root-mean-square value. If when the quantity varies sinusoidally the peak factor is √2.

Peak Forward Voltage : It is the maximum instantaneous voltage applied to a device in the forward direction, *i.e.* in the direction in which the device is designed to pass current with minimum resistance.

Peak Inverse Voltage : It is defined as the maximum instantaneous voltage applied to a device in the reverse direction. In other words it is the direction of maximum resistance.

Peak-Riding Clipper : A limiter which automatically adjusts the value of voltage at which it operates according to the peak value of the pulse train input to the circuit.

Peak Value : It is regarded as the maximum positive or negative value of any alternating quantity, like current or voltage, during a given time interval. The positive and negative values are not necessary equal in magnitude.

It is also called the maximum value of an impulse voltage or current.

Pedestal : It is a pulse waveform in which the amplitude is made constant for a pulse interval large compared to the rise and fall times.

Pentatron : It is thermionic valve that contains five electrodes and is effectively two triodes with a single common cathode, two anodes, and two control grids.

Perfect Crystal : It is a crystal that has a uniform arrangement of atoms throughout and contains no impurities or dislocations; an ideal single crystal.

Pentode : A thermionic value containing five electrodes cathode C, plate P, control grid S_1, Screen grid S_2 and suppressor grid S_2 (*see Fig.*).

The suppressor grid S_3 is connected to the cathode. This helps in repelling the secondary electron back to the plate P from which they are ejected by the primary electrons. The presence of the suppressor grid eliminates undesirable effects of secondary emission. It minimizes the distinction in the characteristics of a valve. Pentodes have all the advantages of tetrodes viz., lawgrid to plate capacitance, high amplification factor and larger power output.

Period : Symbol: T. It is defined as the time required to complete a single cycle of regularly recurring events. The period of an oscillation is related to the frequency, v, and the angular frequency, w, by

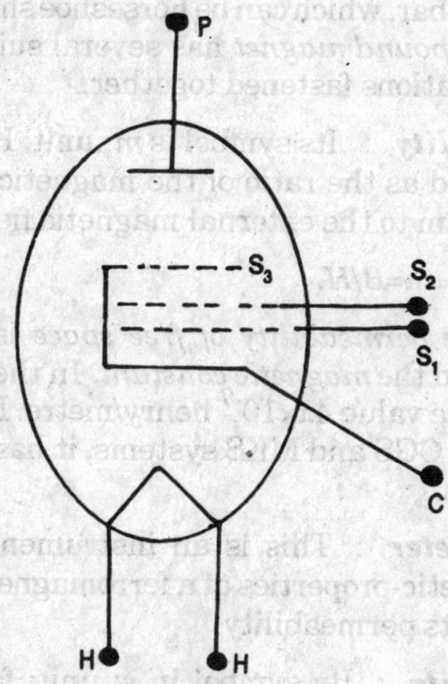

Pentade.

$T = 1/\upsilon = 2\pi/\omega$

Periodic : It denotes only variable quantity that has regularly recurring values with respect to equal increments of some independent variable, such as time. The interval between two successive repetitions is the period.

Periodic Table : The classification of chemical elements, that gives a periodicity of chemical properties when the elements are arranged in order of their atomic numbers. Elements of similar chemical properties occur in the same vertical group in the periodic table The periodic table was used to predict the existence of undetected elements.

Peripheral Devices : The term is used for devices that are connected to a computer, form part of the computer system and whose operation is controlled by the central processing unit of the computer. Terminals, visual display units, printers, magnetic tape units; and disks are examples of peripherals.

Permanent magnet : It is defined as a magnetized sample of a ferro-magnetic material, like steel, which possesses high retentivity and is stable against reasonable handling. It requires a definite demagnetizing flux in order to destroy the residual magnetism. *A simple magnet* has a single bar, which can be horseshoe shaped, of the material. A *compound magnet* has several suitably shaped bars or laminations fastened together.

Permeability : Its symbol is *m*; unit: Henry per meter. It is defined as the ratio of the magnetic flux density, *B*, in a medium to the external magnetic field strength, *H*, *i.e.*

$$\mu = B/H.$$

The *permeability of free space* is designed μ_o and is termed the *magnetic constant*. In the system of SI units it has the value $4\pi \times 10^{-7}$ henry/metre. In other system, such as the CGS and MKS systems, it has been given different values.

Permeameter : This is an instrument that measures the magnetic-properties of a ferromagnetic material, particularly its permeability.

Permittivity : Its symbol in ε; unit: fared per metre. It is defined as the ratio of the electric displacement, *D*, in a dielectric medium to the applied electric field strength, *E*,

$$\varepsilon = D/E$$

It indicates the degree to which the medium can resist the flow of electric charge and is always greater than unity.

The *permittivity of free space* is designed eo and is known as the *electric constant*. It is release to the permeability of free space, μ_o by the equation

$$\varepsilon_o \mu_o = 1/C^2$$

where c is the velocity of light. As μ_o has the value $4\pi \times 10^{-2}$ Henry per meter in the system of SI units,

$$\therefore \varepsilon_o = (1/4pc2) \times 107 \text{ farad/metre}$$

$$8.854 \times 10^{-2} \text{ farad/metre}$$

Persistence : It is defined as the time interval after excitation during which a phosphor continues to emit light particularly the phosphor on the screen of a cathode-ray tube.

It is also defined as a faint luminosity observed after the passage of an electric discharge through certain gases; it can last for several seconds.

Persistor : It is a device, commonly in the form of a miniature bimetallic printed circuit, which depends for its operation on the sharp changes of resistance in a metal as it passes from a state of superconductivity to its normal resistive state. It may be used as a low-temperature storage element or very fast switch.

Perveance : It is defined as the space-charge-limited characteristic between the electrodes in an electron tube. It is a function of the current density, j, and the collector voltage, V, being equal to $j/V^{3/2}$.

Phantom Circuit : This is a circuit which is superimposed on two other circuits, which consist of two pairs of transmission lines, like the pairs of wires of a telephone circuit.

Phase : It is defined as the stage or state of development of a regularly recurring quantity; it is the fraction of the period which has elapsed with respect to a fixed datum point.

Phase Angle : It is defined as the angle between two vectors which show two sinusoidally varying quantities of the same frequency.

Phase Corrector : It is a network which restores the original phase of a waveform that has suffered phase distortion.

Phase Delay : It is defined as the ratio of the inserted phase shift undergone by a periodic quantity to the frequency.

Phase Difference : Its symbol is ϕ. It is defined as the difference in phase between two sinusoidally varying quantities of the same frequency. It may be expressed as an angle-the phase angle-or as a time.

It is also defined as the angle between the reversed secondary vector of an instrument transformer and the corresponding primary vector.

Phase Discriminator : It is defined as a detector circuit which gives an output wave in which the amplitude is a function of the phase of the input.

Phase Inverter : It is a circuit which charges the phase of an input signal by π.

Phase Modulation (p.m) : This is a type of modulation in which the phase of the carrier wave changes about its unmodulated value by an amount proportional to the amplitude of the modulating signal and at the frequency of the signal wave if the amplitude of the carrier is constant.

Phase Sequence : This is defined as an order in which the three phases of a three-phase system reach a maximum potential of given polarity. The normal order in a particular system is termed the *positive sequence;* the reverse order is the *negative sequence.*

Phase Shift : Any charge which takes place into the phase of a periodic quantity or in the phase difference between two or more such quantities. The phase shift can occur as a result of errors introduced by a particular device or circuit or can be deliberately inserted by a phase-shifting network.

Phase Splitter : It is a circuit that has a single input signal and produces two separate outputs with a predetermined phase difference.

Phase Velocity : It is defined as the velocity at which an equiphase surface of a travelling wave is propagated through a medium, or the velocity at which the crests and troughs travel. It is equal to λ/T where λ is the wavelength and T the period of the wave. If the frequency is μ and the number of wavelengths per unit distance is s (the wavenumber) then

$$\lambda/T\text{-}v/\sigma$$

Phase Voltage : It is the voltage in one phase of a polyphase system.

Phasing : It is adjusting the position of the picture transmitted in television or facsimile transmission along the scanning line.

Phosphor-Bronze : Bronze which has at least 0.18% of added phosphorus. The addition of the phosphorus increases the tensile strength, ductility, and shock resistance of the alloy. Phosphor-bronze in strip form has been widely used for galvanometer suspensions and other similar applications.

Photoactor : It is a device which gives the light source to activate a photoconductive photocell when the latter is used as a switch.

Photocathode : It is defined as a cathode that emits electrons as a result of the photoelectric effect.

Photocell : It is a light-electric transducer. The term was originally short for *photoelectric cell,* which is a vacuum diode containing a photocathode and an anode.

The term is now used to designate a *photoconductive cell,* having a semiconductor material sandwiched between two-ohmic contacts.

Photoconductivity : It is regarded as the enhancement of the conductivity of certain semiconductors due to the absorption of electromagnetic radiation. Photoconductivity may result from the action of radiation in the visible portion of the spectrum in some materials.

Photoconverter : It is defined as a photoelectric transducer which gives digital electrical signals from an optical pattern. It consists of any array of photocells each of which produces a current proportional to the intensity of light falling on it. The illumination can be due to an optical image produced on a transparent screen or a pattern resulting from an illuminated photographic mask or cut-out of opaque material.

Photocurrent : It is defined as an electric current produced in a device by the effect of incident electromagnetic radiation.

Photodetachment : The interaction of electromagnetic radiation with a negative ion to detach an electron from the ion and leave a neutral atom or molecule, is called photodetachment.

Photodetector : It is defined as any device which detects or responds to the light energy.

Photodiode : It is a semiconductor diode which produces a significant photocurrent when illuminated. There are two main classes of photodiode : *depletion-layer photodiodes* and *avalanche photo-diodes.* The general symbol is shown in Fig.*a*

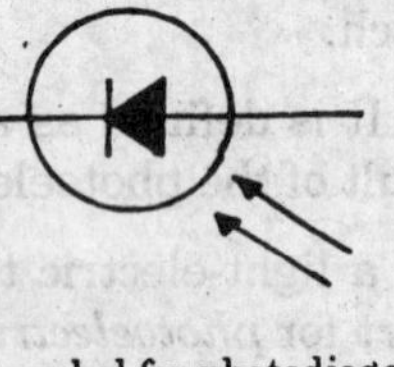

a. General symbol for photodiode

A common form of depletion-layer photodiode contains a reverse-biased *p-n* junction operated below the break-down voltage. When exposed to electromagnetic radiation of suitable frequency excess charge carriers are given out as a result of photoconductivity; these carriers are in the form of electron-hole pairs.

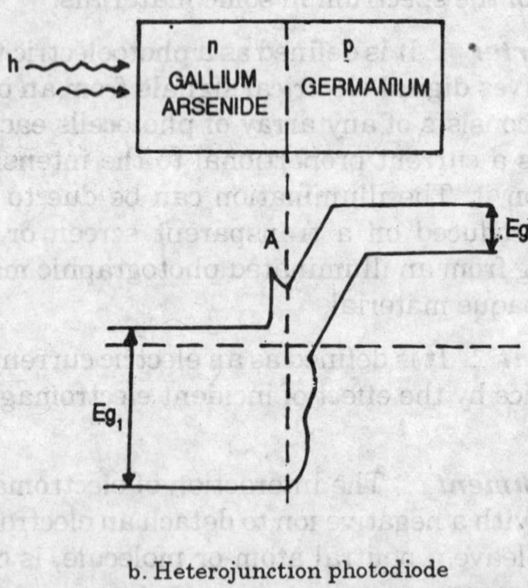

b. Heterojunction photodiode

The heterojunction photodiode is obtained from two semi-conductors of different band gaps (Fig.*b*).

Point-contact photodiodes are obtained with a point-contact junction.

The other class of photodiodes, avalanche photodiodes,

are reverse-biased *p-n* junction diodes that are operated at voltages above the breakdown voltage. Static current characteristics are shown in Fig.*b*. Avalanche photodiodes provide a substantial gain at microwave frequencies.

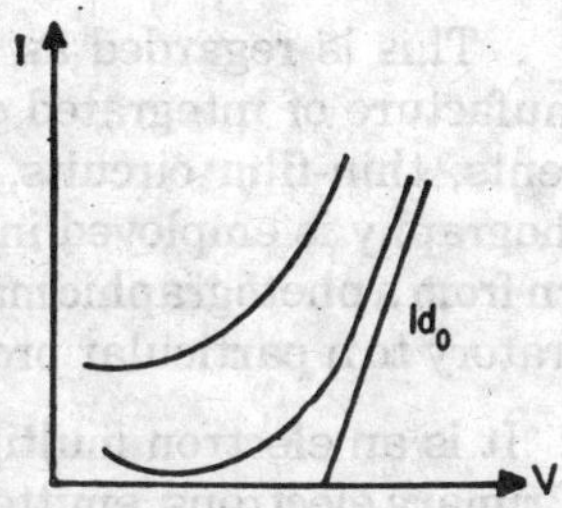

c.h I-V characteristic af an avalanche photodiode.

Photoelectric Alarm : It is a system which uses a photocell used as a switch. The most common form of operation is to arrange for the photocell to be subjected to a constant beam of light.

Photoelectric Constant : It is defined as the ratio of the Planck constant, *h*, to the electron charge, *e*.

Photoelectric Effect : This is an effect, first noticed by Heinrich Hertz, whereby electrons are liberated from matter when it is exposed to electromagnetic radiation of certain energies. In case of solids electrons are only liberated when the frequency of the exciting radiation is greater than a characteristic value-the *photoelectric threshold* of the material. The photoelectric threshold is generally in the mid-ultraviolet region of the electromagnetic spectrum for most solids, although some metals exhibit photoelectric emission with visible or near-ultraviolet radiation. The numbers of electrons rejected from a solid is not dependent on the frequency of the radiation but on the intensity; the maximum velocity however is directly proportional to the frequency.

Photoemission : Emission of electrons form a material as a result of exposure to electromagnetic radiation is known as photo-emission

Photoionization : This is defined as the inonization of an atom or molecule which is due to exposure to electromagnetic radiation. The mechanism of photoionization is the same as theat operating in the photoelectric effect, but in photoionization an electron is ejected from a single atom or molecule, such as occur in the gaseous state.

Photolithography : This is regarded as a technique used during the manufacture of integrated circuits, semiconductor components, thin-film circuits, and printed circuits. Photo-lithography is employed in order to produce a desired pattern from a photographic mask on a substrate material preparatory to a particular processing step.

Photomultiplier : It is an electron multiplier which has a photocathode. Primary electrons, emitted from the photocathode as a result of the photoelectric effect, initiate the cascade. A suitable scintillation crystal is generally employed as the source of illumination for the photocathode in order to provide a sensitive radiation detector or counter.

Photon : It is defined as the quantum of electromagnetic radiation. It may be regarded as an elementary particle of zero mass and having energy hv,where h is the Plank's constant and v the frequency of the radiation. It travels at the velocity of light, c, and has momentum hv/c *or* hl where l is the wavelength of the radiation.

Photoresist : It is a photosensitive organic material used during photolithography.

Photosensitivity : It is a property responding to electromagnetic radiation, particularly in the ultraviolet, visible or infrared portions of the electromagnetic spectrum.

Phototransistor : It is defined as a photodetector which has a bipolar junction transistor operated with the base region floating. The potential of the base region is determined by the number of charge carriers stored in it.

The electromagnetic radiation causes excess electron-hole pairs in the base region. Minority carriers produced in or near to the depletion layers associated with the reverse-biased collector junction are swept into the collector region (*see* diagram); majority carriers remain in the

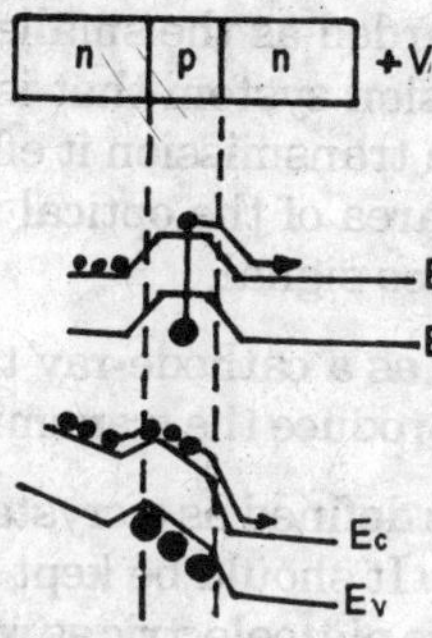

Simplified energy bands and carrier transport in an n pn phototransistor

base region and are stored as excess majority carriers. The emitter junction thus becomes more forward biased.

Phototube : It is an electron tube which has a photosensitive electrode, generally the cathode. A *vacuum photo tube* is evacuated to a sufficiently low pressure such that ionization of the residual gas in the tube does not affect the characteristics. A *gas phototube* is one that contains a gas, like argon, at very low pressure in order to minimize the space charge effect of the photoelectrons in the tube.

Photovoltaic Cell : This is a type of a cell that employs the photo-voltaic effect in order to produce an e.m.f. An example is the solar cell, the basis of which is an unbiased *p-n* junction. Metal-semiconductor junctions are a common form of photo-voltaic cell.

Photovoltaic Effect : The effect taking place in a junction between two dissimilar materials, such as a metal and a semiconductor or two opposite polarity semiconductors, is exposed to electro-magnetic radiation, generally in the range near-ultraviolet to infrared.

Pick-up : It is defined as a transducer which changes information, usually recorded, into electric signals. The term is particularly applied to the electromechanical transducer used to reproduce the signals recorded in the grooves of a gramophone record. There are several types of pick-up in common use.

Pico : Symbol: p. It is a prefix to a unit, denoting a submultiple of 10-12 of that unit; one picofarad is 10-12 farads.

Picture Element : It is regarded as the smallest portion of a picture area in a television system that is resolved by the scanning processes. In transmission it effectively corresponds to the smallest area of the optical image which produces a discernible video signal.

Picture Tube : It is defined as a cathode-ray tube based in a television receiver to reproduce the transmitted picture.

Piezoelectric Crystal : It is defined as a crystal that exhibits the piezoelectric effect. It should be kept in mind that all ferroelectric crystals are piezoelectric as well as certain nonferro-electric crystals and some ceramics.

Piezoelectric Effect : This is regarded as an effect which takes place when certain materials are subject to mechanical stress. An electrical polarization is set-up in the crystal and the faces of the crystal became electrically charged.

The piezoelectric effect is useful as it couples electrical and mechanical energy and thus has many applications for electro-mechanical transducers.

Piezoelectricity : The term is used for electrical signals generated as a result of the piezoelectric effect.

Piezoelectric Oscillator

This is an oscillator which utilizes a piezoelectric crystal in order to determine the frequency. These types of oscillators are very stable.

Pirani Gauge : This is defined as an instrument which measures low pressure by means of the variation of resistance of a conductor with pressure.

Planar Process : The term is used for the most commonly used method of producing junctions during the manufacture of semiconductor devices. A layer of silicon dioxide is thermally grown on the surface of a silicon substrate of the desired conductivity type.

Planck Constant : Its symbol is h. It is defined as a universal constant arising from *Planck's law,* which states that the energy of electromagnetic radiation is due to discrete packets or photons. The energy of each photon is given by *hv*, where *v* is the frequency of the radiation. Plank's law is fundamental to quantum theory. The value of *h* is 6.626196 x 10-34 joule seconds.

Plane Polarized Wave : It is *syn.* to linearly polarized wave. it is defined as an electromagnetic wave in which the vibrations are rectilinear and parallel to a plan that is transverse to the direction of propagation of the wave.

Plane Position Indicator (PPI) : It is a presentation used in a radar receiver in which the signal from the target appears as a bright spot on the screen. The distance and bearing of the target are given by the polar coordinates of the spot with respect to the centre of the screen.

Plasma : It is a defined as a region of ionized gas in an arc discharge tube (*see* gas-discharge tube) which has approximately equal number of electrons and positive ions and provides a conducting path for the arc discharge.

It is also used for a substantially completely ionized gas at extremely high temperature consisting of electrons and atomic nuclei. The material of the sun and other substances capable of thermonuclear reaction are known as plasmas.

Plasma Oscillations : These are oscillations of the ions and electrons in the plasma of gas-discharge tube which take place under certain conditions independently of the conditions in the external circuit.

Plastic-film Capacitor : It is defined as a capacitor the dielectric of which is a plastic film.

Plastics : The term is used for a group of organic materials that are plastic at some stage during their manufacture although stable in use at normal temperatures; they can be shaped by the application of heat and pressure. Plastics are used in electronics, particularly in the packaging of solid-state circuits, components, or devices, as the substrate material for magnetic tape or disk and gramophone records, and as insulating and dielectric material.

Plate : It is defined as an electrode in an electrolytic cell or in a capacitor.

Plateau : The term is regarded as a region of the current-voltage characteristic of an electronic device which shows a substantially constant value of current for a significant voltage range.

Plated Magnetic Wire : A magnetic wire containing surface of ferro-magnetic magnetic material plated on to a non-magnetic core.

Platinum : Its symbole is Pt. A metal, atomic number 78 that is extremely stable and noncorroding and is used as an electrical contact or conductor at high temperatures or when chemical attack is likely.

Plug and Socket : This is a device that enables electrical apparatus to be connected or disconnected from a source of supply or other equipment. It has two separable portions, the 'male' plug and the 'female' socket, with metal contacts that engage each other when connection is effected.

Plug in : It denotes a device which may be rapidly connected or disconnected from a complex electrical or electronic system or apparatus. The device generally consists of a standardized printed-circuit board or integrated-circuit package which may simply inserted in or removed from the main equipment and thus facilitates repair or maintenance of the equipment.

p.m. : It is *abbrev. for* phase, modulation.

p.n. Junction : It is regarded as the region at which two semiconductors of opposite polarity meet, *i.e.* at which a *p-type* and *n-type* semiconductor meet. A simple *p-n* junction is obtained from the same material in which approximately equal doping levels lead to two different conductivity types. A *heterojunction* is formed from two dissimilar materials, such as silicon and germanium. The properties of *p-n* junctions are used in may semiconductors devices, like diodes and transistors.

Pnpn (or p-n-pn) Device : *Syn.* thyristor. It is defined as a semiconductor device which is made of silicon and consists of alternating *p-type* and *n-type* layers and contains at least three *p-n* junctions. The device has a bistable current-voltage characteristic and is used for power-switch-

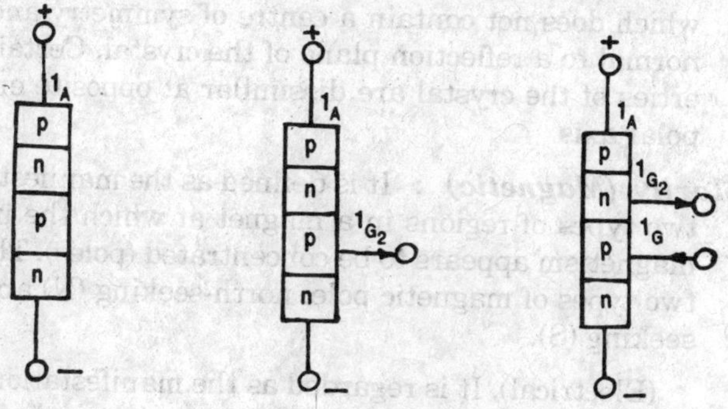

Four-layer diode forward bias   Semi-conductor-controlled rectifier   Semi-conductor controlled switch

(a) Contacts for various pn-pn devices.

ing purposes. A pn pn device with no gate electrodes (Fig *a*) is termed a *four-layer diode (or Shockley diode)* Its switching function is controlled by Pn pn devices are subject to delay in turn-on and turn-off times. The turn-off time is the minimum time interval between removal or reversal of the load current and application of a specified forward blocking voltage without the device turning on.

Point Contact : It is defined as a non-ohmic contact between a pointed metallic wire and the surface of a semiconductor.

Point-Contact Diode : The term is used for a diode rectifier formed by making a point contact between a small metal wire with a sharp point and a semiconductor. The contact may be a simple mechanical contact or a small alloyed contact formed by electrical discharge, as in the gold-bonded diode.

Poisson's Equation : It states that if the electric potential is V at a point (*x*, *y*, *z*) and the charge density is r, then Poisson's equation is expressed as

$$\partial^2V/\partial x^2+\partial^2V/\partial y^2+\partial^2V/\partial z2=4\pi r$$

Polar : It denotes a component, like an electrolytic capacitor, that can only operate normally in one direction of applied voltage or any system, such as a molecule, or any physical property that exhibits asymmetry.

Polar axis : This is defined as a crystal axis of rotation which does not contain a centre of symmetry and is not normal to a reflection plane of the crystal. Certain properties of the crystal are dissimilar at opposite ends of a polar axis.

Polarity. (Magnetic) : It is defined as the manifestation of two types of regions in a magnet at which the inherent magnetism appears to be concentrated (poles). There are two types of magnetic pole: north-seeking (N) and south seeking (S).

(Electrical). It is regarded as the manifestation of positive or negative parameters in an electrical circuit or device.

Polarization : The term is usually used for a phenomenon occurring in a simple electrolytic cell containing dissimilar electrodes.

Pole : It is defined as the terminal or lines to a piece of electrical apparatus, a circuit, or network between which the circuit voltages are applied or produced.

Polychromatic Radiation : The term is used for electromagnetic radiation which has more than one frequency. The term is also applied to particulate radiation when the particles are all the same type but of different energies.

Polyphase System : The term is regarded as an electrical system or apparatus that has two or more alternating supply voltages displaced in phase relative to each other.

Polyphase Transformer : It is defined as a transformer which is employed with a polyphase system. The magnetic circuits required for each of the phase windings usually have portions in common with each other in order to retain the correct voltage.

Port : It is defined as an access point in an electronic circuit, device network, or other apparatus where signals can be input or output.

Post-office Box : It is a box containing a number of resistance coils connected to form three arms of a Wheatstone bridge, the resistance to be measured forming the fourth arm. Each resistance coil is connected between adjacent metal blocks in such a way that it may be shorted out of the circuit by the insertion of a metal plug between the blocks.

Potential Barrier : This is regarded as a region in a field of force in which the potential is of such polarity as to oppose the motion of a particle subject to the field.

Potential Difference. (p.d) : Symbol: ΔV, U. It is defined as the difference in electric potential between two points equal to the line integral of the electric field strength between the points.

Potential Divider : It is regarded as a chain of resistor, inductors, or capacitors arranged in series. It is tapped at one or more points along the chain in order to obtain one or more predetermined fractions of the total voltage across the chain.

Potentiometer : An adjustable resistor connected across a voltage service is known as potentiometer. By moving the contact arm across the resistor any voltage between zero and maximum value of the same can be obtained.

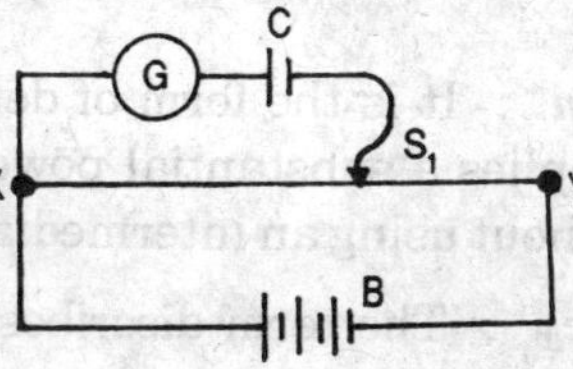

Poteniometer

A typical use is for the measurement of potential difference or e.m.f. by balancing the unknown e.m.f. with that of a standard. For example, to measure the e.m.f. of a cell, C, the cell is arranged as shown in the diagram. The slider, S, is moved along dY until the null position is found on the galvanometer, G. The e.m.f. due to C just balances the potential across XS^1 (of length 11). The cell is replaced by a standard cell, Cs, and the new balance point found at

XS^2 (length 1,). Then

E_c/E_s=11/12

where *Ec and Es* are the e.m.fs. of the unknown and standard cells, respectively.

It is also regarded as any variable resistor, usually wire-wound, used in electronic circuits that has a third moveable contact.

Poulsen arc Converter : It is defined as an arc converter that can generate alternating-current signals of frequencies up to 100 kilohertz.

Powdered-Iron Core : It is defined as a magnetic core which is made of finely divided particles of iron embedded in a plastic or ceramic binding materials. The low dissipation of such a core makes it very useful for high-frequency applications.

Power : Power is the rate of during work. The time used in performing the work in the determining factor is power. Expressed mathematically power=W/T where W represents the warm and T the time needed to do the work.

The electrical unit of power is watt. Large amounts of power an expressed as kilowatts, or mega watts while small amounts are expressed as multiwatts or microwatts.

Power Detection : It is the form of detection in which the detector supplies a substantial power output directly to the local without using an intermediate amplifying stage.

Power Efficiency : The term discribes the ratio of the energy output by a transducer to the energy absorbed by the input circuit under specified operating conditions. The term is especially applied to an electroacoustic transducer, like a loudspeaker. The inverse of the power efficiency is the *power loss* of the device.

Power Factor : It is defined as the ratio of the actual power in watts developed by an a.c. system, as measured by a wattmeter, to the apparent power in volt-amperes, indicated by voltmeter and ammeter readings.

Power Frequency : The term is used to define the frequency at which domestic and industrial mains electricity is supplied and distributed.

Power Pack : It is defined as a device which changes power from an a.c. or d.c. supply, usually the mains, into a form that is suitable for operating electronic devices.

Power Station : It is regarded as a complete assembly having all necessary plant, equipment, and buildings at a suitable site for the conversion of energy of one type, such as nuclear energy into electrical power.

Power Transistor : It is a transistor designed to operate at relatively high values of power or to produce a relatively higher power gain. Power transistors are employed for switching and amplification. They generally require some form of temperature control since the power dissipation in them ranges from 1 watt to 100 watts.

Preamplifier : It is an amplifier which is used in a sound-reproduction system, such as a radio receiver, in order to amplify the received signals before they are input to the main part of the system.

Precision Approach Radar (PAR) : It is defined as a radar system which is employed at an airport to present accurate information about the location of incoming aircraft in the vicinity of the airport and is used as an aid for air traffic control.

Pre-Emphasis : It is regarded as a technique used to improve the signal-to-noise ratio in a radio-communication system that employs frequency modulation or phase modulation.

Pressure-Type Capacitor : It is a capacitor used at high voltage that contains an inert gas under pressure as the dielectric.

Price's Guard Wire : It is a wire conductor used to prevent surface leakage current associated with an insulator from reaching the measuring instrument during determination of insulation resistance, specifically cables.

Primary Electrons : The electrons which impinge on a surface and result in secondary emission of electrons from the surface.

Principal Axis : It is the longest crystallographic axis.

Printed Circuit : It is defined as an electronic circuit, or part of a circuit, in which the conducting interconnection pattern is formed on a board.

Printer : It is regarded as a device which gives printed characters, numbers or symbols on paper from a source of information like a source of experimental data, the output of a computer, information transmitted by a telecommunication system, or a data-bearing medium such as punched cards.

Print-Through : It is a form of distortion arising during magnetic tape recording. It is due to a region of strongly magnetized tape affecting adjacent layers.

Probe : It is defined as an electric lead that connects to a measuring or monitoring circuit, or contains such a circuit and its end or along its length, and that is used for testing purposes.

It is also regarded as a resonant conductor that is inserted into a waveguide or cavity resonator in order to inject or extract energy.

Program : It is a complete set of instructions written in a particular programming language which results in a computer to perform a set of defined operations.

Programme Signal : The term is used for the complex wave containing the information corresponding to the sound information (audio signal) and vision information (video signal) during a specific radio or television broadcast.

Programming Language : It is a language which is used to be mutually comprehensible to computers and humans. A particular digital computer operates in binary notation and all instructions must ultimately be produced in the appropriate binary code, known as *machine code*.

A *high-level programming language* is a language which resemble natural language or mathematical notation more closely than machine code.

Propagation Coefficient : Symbo l: γ, P. it is a complex quantity that expresses the effect of a transmission line on a sinusoidal progressive wave. The propagation coefficient is regarded as a uniform transmission line of infinite length supplied with a sinusoidal current of specified frequency at its sending end.

Propagation Loss : It is the energy loss from a beam of electromagnetic radiation as a result of absorption, scattering, and spreading of the beam.

Proportional Control : It is regarded as a control system which operates by first determining the difference between the actual value of the quantity to be controlled and the desired value, and then applying a correction proportional to this difference.

Proportional Counter : It is defined as a gas-filled radiation detection tube operated in the region of proportionality on the characteristic curve of the detector. In this region gas multiplication takes place in such a manner that the output size is proportional to the number of ion-pairs produced by the initial ionizing event and therefore to the energy of the ionizing radiation.

Protective Gap : It is defined as a surge diverter that is used to protect an electrical system or apparatus from excessively high voltage surges. The term does not apply to any series resistor that may be present in the earth connection.

Protective Relay : It is defined as a relay which is employed to protect electrical apparatus against overload and fault conditions.

Proton : It is an elementary particle having a positive charge equal to that of the electron, which is about 1836 times than the electron. The proton forms the nucleus of the hydrogen atom and is a constituent part of all atomic nuclei. The number of protons in an atomic

Proximity Effect : It is regarded as an effect observed when two or ore conductors carrying alternating current are placed close to one another.

Psophometer : It is an instrument which given the amount of noise in a transmission system, particularly a telephone system. It is also defined as a device for measuring noise power through a weighting network in order to produce objective results that approximately parallel subjective results with human observers.

p-Type Semiconductor : It is a semiconductor having a higher density of mobile holes than of condition electrons.

Public Address System (PA) : It is regarded as a complete reproducing system which is employed to render sounds audible to a large gathering of people.

Pulling : It is defined as a change of frequency observed in an electronic oscillator when it is coupled to a circuit containing another independent oscillation.

Pulsating Current : It is defined as a current that exhibits regularly recurring variations of magnitude. The term implies that the current is unidirectional.

Pulse : It is defined as a single transient disturbance manifest as an isolated wave, one of a series of transient disturbances recurring at regular intervals, or a short train of high frequency waves. A pulse consists of a voltage or current which increases from zero (or a constant value), to a maximum and then decreases to zero (or the constant value), both in a comparatively short time. The zero or constant value in the absence of the pulse is called the *base level.* A pulse is described according to its geometrical shape when the instantaneous value is plotted as a function of time : it can be rectangular, square, triangular, etc. Unless otherwise specified a pulse is assumed to the *rectangular.*

Pulse Communications : These are defined as the telecommunications involving the transmission of information by means of pulse modulation.

Pulse Discriminator : It is a discriminator which selects and responds only to a pulse with a particular characteristic of amplitude, period, etc., and which is employed in pulse operation.

Pulse-forming line : It is an artificial line that has inductances and capacitors in series and is used to produce fast high voltage pulses in radar.

Pulse Generator : It is regarded as an electronic circuit or device which produces voltage or current pulses of a desired waveform. It can be designed to produce either single pulses or a pulse train.

Pulse Modulation : This is a type of modulation in which pulses are employed to modulate the carrier wave or in which a pulse train is used as the carrier (the *pulse carrier*).

Pulse modulation is commonly used for time division multiplexing.

Pulse Operation : The term is used for any method of operation of an electronic circuit or device which changes electrical energy in the form of pulses.

Pulse Regeneration : It is regarded as the process of restoring a pulse or pulse train to the original form, timing, and magnitude.

Pulse Shaper : The term is used for any circuit or device which is employed to alter any of the characteristics of a pulse or pulse train.

Pulse Spacing : It is *syn.* for pulse separation; pulse interval; pulse repetition period.

Pulse Width : It is *syn.* for pulse duration; pulse length. See pulse.

Pulse-Width Modulation : It is *syn.* for pulse-duration modulation.

Punched Card : It is a card on which information can be carried in the form of holes punched in columns. There are generally 80 columns per card. Each column has a combination of holes punched in it which represents a specific character, number or symbol. Punched cards were used as original storage medium for digital computers and other devices. The use of cards for the input of information to computers is currently declining despite the ease with which small decks of cards may be hand-edited.

Puncture Voltage : It is defined as the value of voltage which results in an insulator to be punctured when it is subjected to a gradually increasing voltage.

Purple Plugue : It is defined as the formation of purple-coloured areas on the bonds of a silicon integrated circuit.

Push-Pull operation : The term is used for the use of two matched devices so that they operate with a 180° phase difference. The output circuits combine the separate outputs in phase (Fig.*a*). One common means of achieving the desired 190° phase shift in the inputs is a transformer-coupled input circuit Complementary transistors may also be used in which case no phase shift is required in the inputs.

Push-pull circuits are generally employed for class A and class b amplification and are then termed *push-pull amplifiers.*

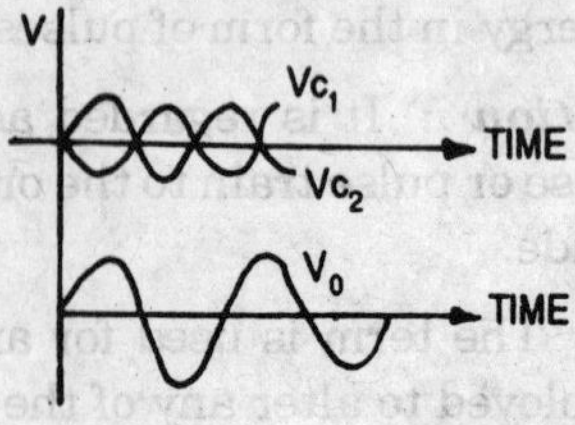

a. Output of transformer-coupled push-pull operation.

Pyroelectricity : The term is used for the development of opposite electric charges at the ends of the polar axes in certain crystals when subjected to a temperature gradient.

Q

Q Band: : It is defined as a band of microwave frequencies ranging from 36.0 to 46.0 gigahertz.

Q Factor : It is a ratio of the circuit reactance to circuit resistance and is a measure of the efficiency storage in a reactive circuit mathematically $Q=\frac{X}{R}$ where *Q* depends upon frequency since reactance appears in the equation.

Quadrant Electrometer : It is an electrometer which is made of a flat cylindrical metal box formed from individually isolated quadrantal segments and containing a light foil-covered vane that is supported by a quartz fibre and is free to move. Opposite quadrants Opposite quadrants are connected together. The instrument is generally arranged in such a way the vane hangs symmetrically within the quadrants when it and both pairs of quadrants are at zero potential.

Quadraphony : It is regarded as a sound reproduction system which is an extension of stereophonic sound reproduction and uses four channels, feeding four separate loudspeaker. The sound may be recorded either using four coincident directional micro-phones at right angles to each other and placed at the position of the listener, or using separated microphones having outputs which are divided between the channels and combined in the correct proportions to achieve the desired effect.

Quadrature : The term is defined as two period quantities having the same frequency and waveform when the phase difference between them is π/2 (90°).

Quadripole : It is *syn.* to four-terminal networks; two-port network. It is defined as a network that has only four terminals, *i.e. a* pair of input terminals and a pair of output terminals. The behaviour of a quadripole is generally described by the impedances presented at its terminals at specified frequencies.

A very common arrangement, used particularly for attenuators and filters, is a ladder network consisting of a number of series and shunt impedances (Fig.a).

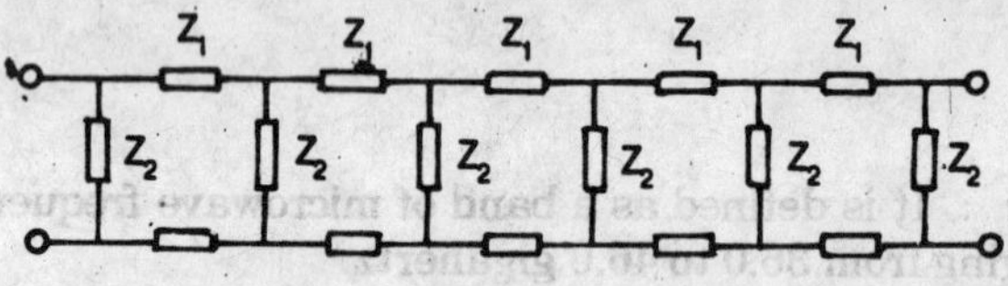

a. Generalised ladder filter

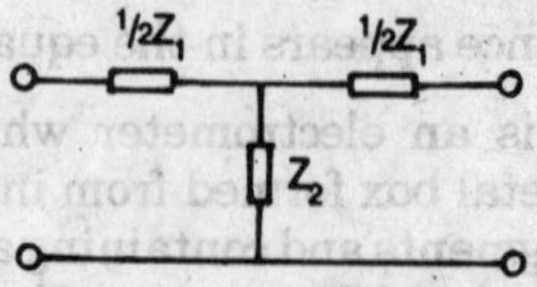

b. T-section

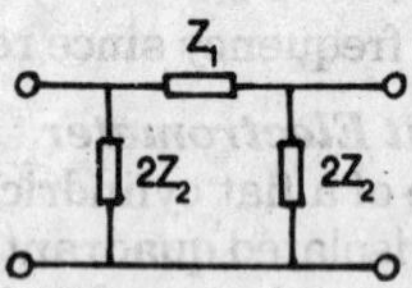

c π-section

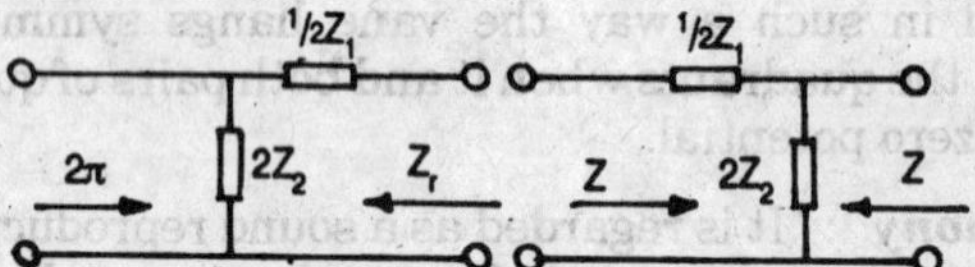

d. L-sections

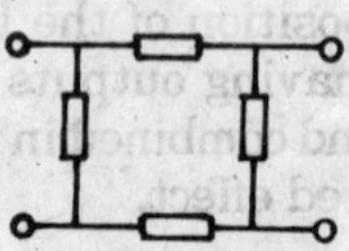

e O-NETWORK

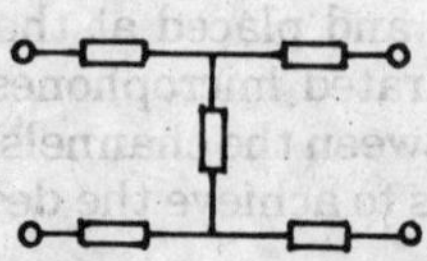

f H-NETWORK

The Ladder filter shown in Fig.a may be analysed as a series of *T-Sections* (Fig.*b*) terminated in the iterative

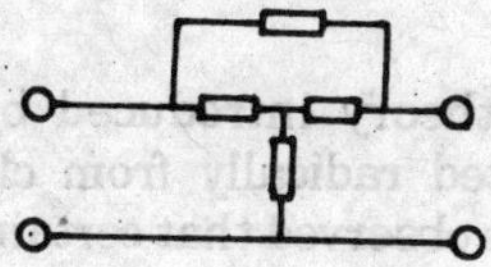

G Bridged-T network

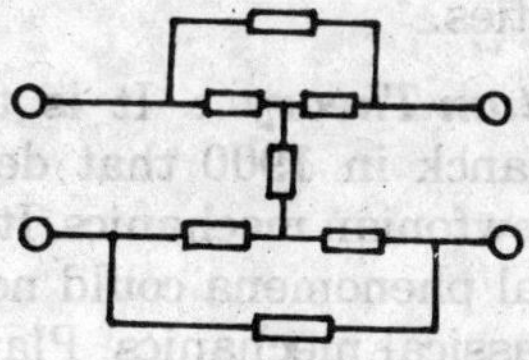

f Bridged-H network

impedance Zp. The same network may be regarded as a serious of p *sections* (Fig. *c*) which must be terminated in the iterative impedance Zp.

Comparison of the two indicates that the half-section or L-sections (Fig.d) acts as an impedance transformer from Zp to ZT.

In general Z_1 and Z_2 and hence ZT and Zp will be dependent on frequency. If the product ZTZp is substantially independent of frequency the network is a *constant R-network.*

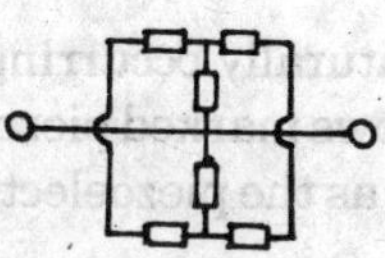

i Twin-T network

Other arrangements of elements used to form ladder filters or attenuation are the *O-networks, H-networks, bridged T, bridged-H, and twin-T networks* (Figs. *e, f, g, h, and i.*)

Quality Control : The term is used for a method of inspection used during the mass production of electronic components, circuits, devices or other apparatus in order to ensure that the finished products conforms to specifications. it generally involves exhaustive checking of random samples at every stage during manufacture.

Quantization : It is regarded as a method of producing a set of discrete or quantized values that represents a continuous waveform.

Quantum Numbers : It is a set of number that are used to label the various possible values of certain physical properties.

Quantum Theory : It is the theory introduced by Max Planck in 1900 that departed radically from classical Newtonian mechanics. It was observed that certain physical phenomena could not be satisfactorily explained by classical mechanics. Planck postulated that radiant energy could be radiated, transported, or absorbed in the form of discrete indivisible packets known as *quanta.*

Physical quantities which are restricted to a number of discrete values by quantum theory are said to be *quantized.* The changes of energy that correspond to a quantum, or an integral number of quanta, are small and only significant on an atomic scale.

Quantum Yield : It is *syn.* to quantum efficiency. The number of reactions of a particular type induced per photon of absorbed electromagnetic radiation.

Quarter-Wavelength Line : It is defined as the transmission line of length equal to one quarter of the wavelength of the fundamental frequency.

Quartz : It is the naturally occurring crystalline silicon dioxide (SiO_2). It shows marked piezoelectric properties and is frequently used as the piezoelectric crystal in piezoelectric oscillator.

Quartz may be readily drawn into extremely fine uniform filaments.

Quasi-Bistable Circuit : It is a multivibrator which is as a result of operating by a trigger (*i.e.*, it is clocked) rather than being allowed to run freely.

Quench : It is capacitor, resistor, or combination of the two, which is employed in parallel with a contact to an inductive circuit and protects spark discharge across the contact when the current ceases.

Quick-Break Switch : A manually operated switch which is designed to operate at a speed independent of that at which the operating handle is moved.

Quiescent Component : It is defined as a component of an electronic circuit which at a specific instant, is not in operation but at a short time following the specified instant will become operative.

Quiescent Current : It is defined as the current which flows in any circuit under specified normal operating conditions but in the absence of an applied signal.

Quiescent Point : It is regarded as the region on the characteristic curve of an active device, such as transistor, during which the device is not operating.

R

Radar : It means radio detection and ranging. It is essentially an echo-ranging system in wireless first similar to echo-ranging in sound. Some types of radar systems in use are :

Pulse radar system

Continuous-wave systems

Doppler radar

Frequency modulated radar

Volumetric radar.

The *radar range* is the maximum distance at which a particular radar system is effective in detecting a target. It is usually defined as the distance at which a designated target is distinguished for at least 50% of the transmitted pulses.

Radar systems are employed for the detection and control of aircraft, guiding of ships in fog, and for locating distant storm centres when an echo is produced by the associated heavy rainfall. Radar is used in astronomy and has an extremely wide range of military applications.

Radar Indicator : It is a cathode-ray tube that provides a visual display of the return echoes in a radar system.

Radar-beam Tubes : It is defined as a electron tube in which the electron beam travels in a radial path from a central cathode to one of a set of circumferential anode.

Radiation : It is a form of energy which is propagated as waves or streams of charged particles.

Radiation Pattern : The term is used for a diagram which represents the distribution of radiation in space from any source, like transmitting aerial, or conversely shows the effectiveness of reception of a receiving aerial.

A typical radiation pattern has one or more *lobes* which represents regions of enhanced response of an aerial in a horizontal or vertical plan. The *major lobs* is lobe which contains the region of maximum radiation intensity or maximum sensitivity of detection of an aerial. it generally points forwards along the direction of propagation of the radiation. A *back lobe* is a lobe which points in the reverse direction to the direction of propagation. A typical radiation pattern is a polar figure of eight (*see* diagram). Point A (r.q.) on the diagram indicates the sensitivity of the aerial shown by *r* at an angle *q* to the direction of propagation of the radiation.

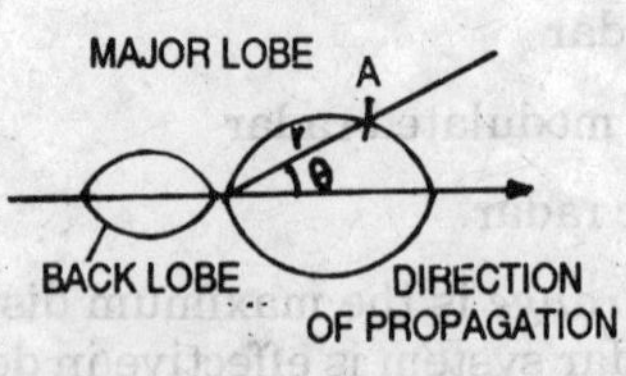

Figure-of-eight radiation pattern

Radio : It is the use of electromagnetic radiation of frequency within the radiofrequency potion of the electromagnetic spectrum for the transmission and reception of electrical impulses or signals without connecting wires or waveguides. It is also the process of transmitting or receiving such signals.

Radioastronomy : It is a science used for study of astronomical bodies and events by means of the radio signals associated with them.

Radio Compass : It is a navigational aid carried on board aircraft and ships. It consists of a radio receiver together with a directive aerial and is essentially a direction finder.

Radiofrequency : It is a frequency of electromagnetic radiation or alternating currents in the range three kilohertz to 300 gigahertz.

Radiofrequency heating : It is defined as dielectric heating or induction heating which is carried out using an alternating field of frequency greater than about 25 kilohertz.

Radio Noise : The term is used for any undesired sound or distortion appearing at the loudspeaker of a radio receiver.

A very common form of noise is *mains hum,* which is due to harmonics of the mains frequency being detected and amplified by the radio receiver, and results in a humming noise. It tends to worsen as the components in the receiver age.

Radio Pill : It is a small capsule which has a miniature radio transmitter and may be swallowed. Miniature transducers respond to conditions.

Radio Receiver : It is a device which changes transmitted radiowaves into audible signals. A simple receiver contains an aerial, a tuner that may be adjusted to the desired carrier frequency, a preamplifier, detector, audiofrequency amplifier, and a loudspeaker. Modern radio receivers generally use superheterodyne reception in order to improve the signal-to-noise ratio of the output.

Radio Set : It is a defined as a combined radio transmitter and receiver which is used by amateur radio operators, in aircrafts or in ships.

Radiospectroscope : It is a device that analyses and displays the total radiofrequency signals received at an aerial.

Radiotelescope : It is defined as a telescope used in radioastronomy which finds extraterrestrial radio signals.

There are two basic forms of radiotelescope. The first consists of large steerable parabolic reflector, or dish, which can be directed at a desired region of the sky. Radiowaves obtained within a pencil-beam emanating from the portion of sky being studied are received and amplified.

The second basic type is the *radio interferometer*.

Radiowave : It is defined as a electromagnetic wave that has a frequency lying in the radiofrequency range.

Radio Window : It is the range of radiofrequencies which are not reflected by the ionosphere but pass straight through it. The radio window extends from about 50 gigahertz to about 15 megahertz (approximate wavelength range : 6 mm to 20m).

Random-access Memory (RAM) : It is *syn.* to direct-access memory. It is a memory designed in such a way that location of the data stored in it is independent of the content and any location in the memory may be directly accessed without having to work through from the beginning. The access time to a given memory location of that type a memory is substantially constant for all memory locations.

RAMs are formed either as moving magnetic surface devices, such as disk or drum, or as electronic memories, such as solid-state memory.

Range : It is defined as maximum distance from a radio or television transmiter at which reception of the signal is possible.

Range Tracking : It is regarded as operation of a radar system when viewing a moving target so that the transmit-receive switch is automatically adjusted to switch to the receive mode at the correct instant for receiption of the return echo.

Rating : It is regarded as stipulating or the stipulation of operating conditions for a machine, transformer or other device or circuit and stating the performance limitations of such equipment.

R-C. : It is *abbre. for* resistance-capacitance. It is used as a prefix to describe circuits or devices which depend critically for their operation on their resistance and capacitance or that employ resistance-capacitance coupling.

R.C.Network : It is short form of resistance-capacitance network.

Reactance : Sumbol: X; unit: ohm. The part of the total impedance of a circuit not due to pure resistance. It is the imaginary part of the complex impedance, Z.

$$Z=R+iX$$

where R is the resistance and i is equal to $\sqrt{1}$.

Reactance Chart : It is a chart which is presented in a form that enables the user to read directly the reactance of any given capacitor or inductor at a specified frequency, and conversely to deduce to capacitance or inductance of a given reactance at a particular frequency.

Reactance Transformer : It is a device which has pure reactances arranged in a suitable circuit and generally employed for impedance matching at radiofrequencies.

Reactivation : It is a process used with a thoriated tungsten filament cathode in order to improve the emission of electrons from the surface.

Reactive Current : It is *syn.* to reactive component, wattless component, idle component, quadrature component of the current.

Reactive Factor : It is depend as the ratio of the reactive volt-amperes of any load, circuit, or device to the total volt-amperes.

Reactive Load : It is load in which the current and voltage at the terminals are out of phase with each other.

Reactive Voltage : It is *syn.* to reactive component, wattless component, idle component, quadrature component of the voltage.

Reactive Volt-amperes : It is *syn.* to reactive component wattless component, idle component, quadrature component of the volt-amperes.

Reactor : This is another term for inductor. Generally reactors are made with iron cores and many have several windings arranged in such a way that their fields produced by each winding aid one an other *i.e.*, their north poles point in the same direction through the core material.

Read : The term is used to generate an output signal from a charge storage tube that corresponds to the information stored in the tube.

It is also defined as to remove information from a computor storage device or memory.

Read-around Number : It is regarded as the number of times as particular location of a charge storage tube may be consulted before information is lost from surrounding locations due to spill.

Read-only-memory. (ROM) : *Syn.* fixed store. A memory which retains information peremanently and in which the stored information cannot be charged by a program or normal operation of computer.

Information can be placed in the storage array during manufacture of the memory-so called *'hardwiring'* and the two possible binary states of each memory location are determined by the physical construction of the device.

Read-out Pulse : It is defined as a pulse applied to a word line of a random-access memory in order to enable a particular storage location on that line to become available to the sense amplifier connected to the appropriate bit line.

Read-write Memory : It is regarded as a memory used in computing in which the stores information can be readily altered.

Real Time Operation : The working of a computor, in particular an analog computer, during the actual time in which a physical process occurs.

Receiver : It is defined as a part of telecommunication system that converts transmitted waves into a desired form of output.

Reciprocal Theorem : It states that if an electromotive force, E, applied at a point in a network produces a current, I, at a second point, and if the same e.m.f. is applied at the second point it will produce the same current at the original point.

Recombination Processes : The different processes by which excess electrons and hols in a semiconductor recombine and ten to restore the system to the thermal equilibrium condition represented by the expression

$pn=n_i2$

p is the number of holes *n* the number of electrons, and ***ni*** the number of holes or electrons in the intrinsic semiconductor at the same temperature.

The basic recombination process are *band-to-band recommbination.*

Recombination Rate : It is defined as the rate at which recombination of electrons and holes in a semiconductor takes place.

Record : It is regarded as a process of making a permanent or semipermanent record of electrical data of various types particularly recording of sound.

Recording Channel : It is defined as one of the independent recorders in a recording system that uses two or more recorders, as in stereophonic recording.

It is also defined as an independent track on a recording medium.

Recording of Sound : It is the process of producing a permanent or semi-permanent record of sounds which may be employed in suitable replay or reproducing apparatus in order to reproduce the original sounds. The three main methods of sound recording the *electromechanical sound recording, photosensitive recording (or sound-on-film),* and magnetic recording.

Gramophone records are mostly used in commercial mass-production for home reproduction or for broadcast entertainment.

Photosensitive or sound-on-film recording is used mostly in the cinema or for broadcast entertainment when film is used.

Megnetic recording is mostly used for a variety of modern applications. The overall performance is as good as or better than, the above two systems and it is very convenient in use. It is replacing gramophone recording for many functions.

Rectifier : It is defined as a device which passes current only in the forward direction and can therefore be used as an a.c. to d.c. converter.

Rectifier Instrument : It is a d.c. instrument which can be made suitable for a.c. measurement by using a rectifier to convert the alternating current to be measured into a unidirectional current. A common arrangement is a bridge circuit formed from four metal or semiconductor diodes and a moving coil instrument, M is shown in Fig. The indicated value on the instrument is generally the root mean-square value of a sinusoidal a.c. input waveform; rectifier instruments are subject to waveform error in case used with non sinusoidal input waveforms.

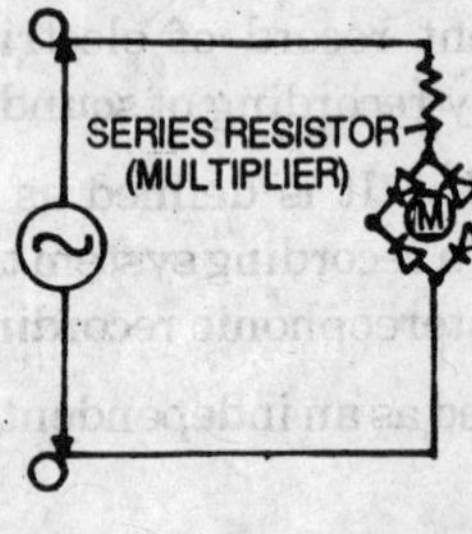

Rectifier Instrument used as an a.c. voltmeter

Rectifier Voltmeter : It is a voltmeter containing a rectifier circuit in the input which converts alternating voltage into an essentially unidirectional voltage, which is then measured.

Recurrent-Surge oscilloscope : This is an instrument used to investigate electrical surges.

Redundancy : It is defined as the fraction of the information in a transmission system which may be eliminated without loss of essential information. The excess information is generally considered to allow for loss in the transmission system.

Reflected Plower : It is defined as the power which is returned from the load back to a generator.

Reflection Coefficient : It is *syn.* for return-current coefficient. It is the vector ratio of the *return current,* I_R, *to the incident current* I_o,

when a transmission line is incorrectly terminated with an impedance, ZR, not equal to the characteristic impedance, Zo, of the line.

Reflection Error : It is an error arising in radio navigation system, direction finding and radar systems as a result of undesired reflections of the transmitted energy.

Reflex Circuit : It is defined as a circuit which is employed to amplify a signal at one frequency and also after it has been converted to a second frequency.

Refresh : The term is used to restore the condition of a memory cell in a computer to its original state in order to maintain he integrity of the information stores in it.

Regenerative Receiver : It is defined as a radio receiver used with amplitude-modulated radiowaves in which positive feedback is employed in order to increase the sensitivity and selectivity of the receiver by reducing the damping.

Register : It is *syn.* for accumulator. One of a number of word-sized locations in the central processing unit of a computer having the arithmetic and logical operations required by a program are performed on data obtained from memory, input/output devices, or other registers.

Regulator : It is an electronic device that is used to maintain the voltage or current constant at a given point in a circuit or to vary it in a controller manner. Regulators are employed to control the output of a load or to regulate the voltage or current of an electronic device, despite fluctuations in the current conditions, particularly variations in the supply voltage.

Reignition Voltage : It is defined as the voltage required to re-establish the discharge in a gas-discharge tube after the tube has ceased conducting.

Rejection Band : It is a band of frequencies which are attenuated or suppressed by a transducer or filter.

Relaxation Oscillator : It is an oscillator having one or more voltages or currents change suddenly at least once during each cycle. The circuit is arranged in such a manner that during each cycle energy is stored in and then discharged from a reactive element (*e.g.*,a capacitor or inductance), the two processes occupying very different time intervals. An oscillator of this type has an asymmetrical output waveform that is far from being sinusoidal.

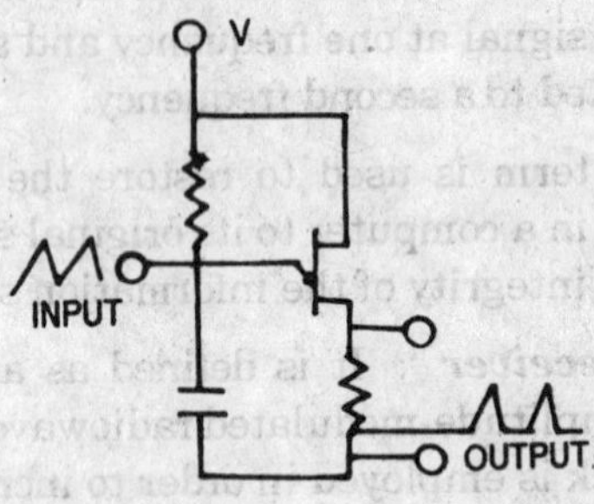

Uni junction transistor relaxation oscillator.

Relaxation Time : Symbol The time interval during which the dielectric polarization of a point in a dielectric falls to l/e of its original value due to the electric conductivity of the dielectric.

Relay : It is an electrical device in which one electrical phenomenon (current, voltage, etc.) controls the switching on or off of an independent electrical phenomenon. There are many types of relay, most of which are either *electromagnetic or solid-state relays.*

Electromagnetic relay is formed when a coil wound on a soft-iron core attracts a pivoted armature that operates contacts. There are various types of armature relays as mentioned today following the principle of electromagnetic relays:

(a) Differential Relay(b) Polarized Relay

(c) Reed Relay(d) Diaphragm Relay

Reliability : It is the ability of any device, component, or circuit to perform a required function under stated conditions for a stated period of time.

Reluctance : In magnetic circuit the opposition to the passage of flux through the system is called reluctance, and it is the magnetic analog of electrical resistance. Symbol : R; unit: henry1. -The ratio of the magneto-motive force, Fm, to the total magnetic flux, f

$$R=Fm/\phi$$

Reluctivity : The receiprocal of magnetic permeability.

Remanence : It is *syn.* to retentivity; residual magnetism.

Remote Control : It is a system of electrical or electronic control that has the controlling equipment located at a distance from the controlled equipment.

Repeater : It is a device which receives signals in one circuit and automatically delivers corresponding signals to one or more other circuits.

Repeating Coil : It is an audiofrequency transformer which employed to couple two sections of telephone line.

Reproduction of Sound : The reproduction f sound information from a source of audiofrequency electrical signals. A complete sound reproduction system has the original source of audio information, preamplifier and control circuits, audio-frequency power amplifier(s), and loudspeaker(s). The source of sound may be a gramophone record, magnetic tape, a breadcast ransmission, or a sound-on-film recording.

Monophonic sound reproduction uses only a single audio-frequency channel.

Stereophonic signals require at least two microphones.

(Hi-Fi) High Fidility is a sound-reproduction system that is composed of high-quality expensive parts and reproduces the original audio information faithfully and with very low noise levels.

Reset : The term is used to mean to restore an electrical or electronic device or apparatus to its original state following operation of the equipment.

Residual Charge : The portion of the charge stored in a capacitor that is retained when the capacitor is discharged rapidly and may be withdrawn from it subsequently.

Residual Current : The current which flows for a short time in the external circuit of an active electronic device after the power supply to the device has been switched off.

Residual Resistance : The inherent resistance of a conductor which is independent of temperature variations. It is generally regarded as irregularities in the molecular structure of the material.

Resistance : Its symbol is R; unit: ohm. It is the property of matter that creates opposition to the passage of electric current through a material. It converts electrical energy into heat energy. It is the ratio of the applied potential difference across a conductor to the current flowing through when the current is an alternating current the resistance is the real part-the *resistive component* of the electrical impedance, Z:

$$Z+R+iX$$

where i is equal to $\sqrt{}^{-1}$ and X is the reactance.

Resistance Brazing : A furnace that utilizes the heating effect of a current in order to provide the heat source.

Resistance Gauge : It is a guage which is employed to measure high fluid pressures by measuring the change in electrical resistance of a sample of manganin or mercury when the sample is subjected to the pressure.

Resistance Lamp : This is an electric lamp which is employed to limit the current flowing in a circuit.

Resistance Thermometer : This is an electrical thermometer which employs the change in electrical resistance with temperature of a wire to measure the temperature of its surroundings.

Resistance Welding : It is defined as an electrical welding in which the heating effect of a current is utilized to provide the heat source. An electric current is passed in controlled short pulses across the contact surface between the components to be welded and local melting of the metal occurs.

Resistance Wire : It is a wire made from a material, such as nichrome or constantan, having a high resistivity and low temperature coefficient of resistance.

Resistivity : Symbol: *r*, unit: ohm metres. This is a intrinsic property of a material equal to the resistance per metre of material with cross sectional area of one square metre and is given by

$$\rho = RA/L$$

where R is the resistance, A the cross-sectional area, and L the length.

Resistivity is the reciprocal of conductivity; the lower the resistivity of a material the better conductor it is. Materials can be classified as conductors, semiconductors; or insulators according to their resistivities.

Ressistor : It is *syn.* to resistance. Any device that is used to introduce additional resistance into an electrical circuit is known as a resistor. These are made in many physical and electrical sizes. The type of resistor chosen depends on the particular application for which it is designed. The three main types of resistance element are *carbon, wire-wound,* and film resistors. These can be produced as fixed-value resistors. These can be produced as fixed-value resistors or adjustable resistors.

Resistor-Transistor Logic (RTL) : The term is applied to a family of integrated logic circuits having the input is through a resistor into the base of an inverting transistor. The basic NOR gate is shown. RTL is of interest since it was the first integrated logic family to be developed.

Resonance : This is regarded as a condition existing when an oscillatory circuit responds with maximum amplitude

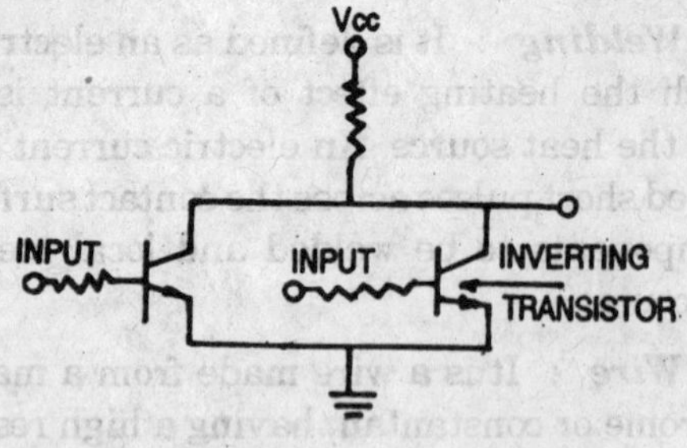

RTL twi-input NOR GATE

to a periodic driving force in such a way that a relatively small amplitude of the driving force produces a large amplitude of oscillation.

Resonance Bridge : It is defined as a four-arm bridge having a tuned circuit in one of the arms. The bridge will only balance at the resonant frequency of the circuit and at balance the frequency and resistance of the circuit are given by the expression

$$\omega^2LC=1$$

$$Rx=RsRa/Rb$$

where ω is the angular frequency.

Resonant Circuit : It is described as a circuit that contains both inductance and capacitance arranged in such a way that the circuit is capable of resonance. The frequency at which resonance occurs-the resonant frequency-depends on the value on the circuit elements and their arrangement.

Resonant Frequency : It is the frequency at which resonance occurs in a particular circuit or network. Resonance takes place in a circuit containing both capacitance and inductance if the imaginary component of the complex combined impedance of the circuit is zero, in other words then the supply current and voltage are in phase and the circuit has unit power factor.

For a series resonant circuit (Fig.a) we have

$$Z=R+iwL-i/wC$$

where R=ohmic resistance, w=the angular velocity, l=inductance, C=capacitance and i=À-1. The resonance condition is given by

$$wol = \frac{1}{woC}$$

or $\omega o=1/\sqrt{(LC)}$

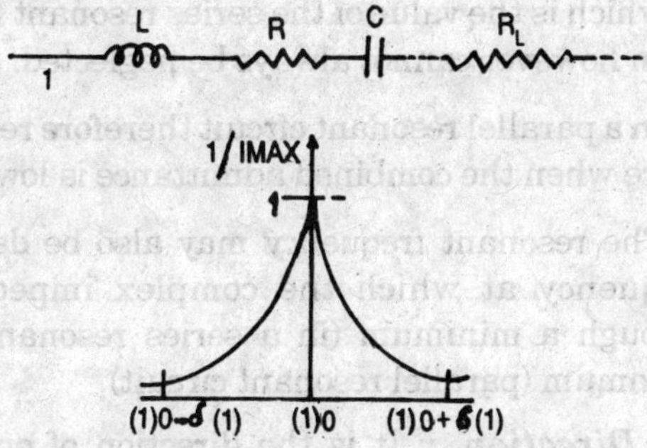

a. Series resonant circuit and frequency response

In a series resonant circuit resonance takes place when the combined impedance is purely resistive and is a minimum.

In the case of a parallel resonant circuit, in which the circuit elements are in parallel (Fig.*b*), it is convenient to consider the combined admittance, Y, of the circuit (Y=Z-

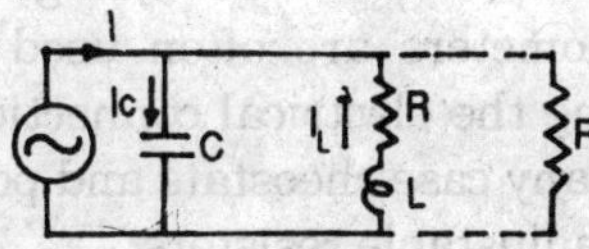

b. Parallel resonant circuit

1), given by the equation

$$Y=i\omega C+(R-i\omega L)/R^2+\omega^2L2)$$

The resonance condition is given by

$R^2+\omega_o 2L^2=L/C$

or $w_o 2=[1-(R2C/L)][1/LC]$

As the term R^2C/L is gener lly very small this approximates to

$wo=1/\sqrt{(LC)}$

which is the value of the series resonant frequency; the term however cannot always be neglected.

In a parallel resonant circuit therefore resonance takes place when the combined admittance is low.

The resonant frequency may also be defined as that frequency at which the complex impedance passes through a minimum (in a series resonan circuit) or a maximum (parallel resonant circuit).

Reverse Direction : It is the direction of operation of an electrical or electronic device which shows larger resistance. A voltage applied in the reverse direction is a *reverse bias* and the current flowing is the *reverse current*.

The *reverse recovery time* of a semiconductor diode, is the time interval between instantaneous switching from a forward bias to a reverse bias and the reverse current reaching the saturation value.

Rheostat : It is a similar to a potentiometer. But potentiometers are generally connected across a voltage source while rheostats appear in series. Thus a rheostat is more of controller of current than the voltage in a given circuit. Small potentiometers are often used as rheostats by merely ignoring the electrical connections at one end of the device. In any case rheostats and potentiometers are both variable adjustable resistors.

Ribbon Microphone : A microphone having a very thin ribbon of aluminium alloy a few milli-meters wide loosely fixed in a strong magnetic flux density parallel to the plane of the strip.

Richardson's Equation : It is *syn.* to Dushman's equation; Richardson-Dushman equation. The fundamental equation of thermionic emission which gives the relation between the number of electrons emitted from a surface to the temperature of the body and expressed as:

$$j=AT^2\exp(-b/T)$$

where *j* is the emitted current density, *A* and *b* are constants, and *T* is the thermodynamic temperature. *A* is a function of the surface state of the material:*b* equals *l*/*k*,where l is the work function and k the Boltzmann constant.

Rieke Diagram : A diagram which indicates the performance data of a microwave oscillator, particularly a magnetron or klystron. It indicates the variation of power output, anode voltage, efficiency, and frequency with changes in the ratio of the maximum and minimum voltages of the standing wave and phase angel in the load at fixed typical operating conditions.

It should be kept in mind that the Rieke diagram is plotted on polar coordinates.

Ringing : It refers to undesired unwanted low-frequency resonant tone which takes place in a radio receiver due to low frequency oscillations produced in the receiver by the received radiofrequency waves.

Ring Winding : Or toroidal winding; Gramme winding. A process of fabricating a coil, particularly for a winding in an electrical machine, in which the coil is wound on an annular core.

Ripple : It means an alternating-current component superimposed on a direct-current component resulting in variations in the instantaneous value of a unidirectional current or voltage. The term is mostly applied to the output of a rectifier.

The magnitude of the ripple is obtained by the ratio of the root-mean-square value of the a.c. component to the mean value of the total and is usually expressed as a percentage. This is termed as the *ripple factor*.

Root-Mean-Square (r.m.s) Valve or Effective Value; Virtual Value : The square root of the mean value of the squares of the instantaneous values of a periodically changing quantity averaged over one complete cycle. In the case of a sinusoi dally varying quantity the r.m.s. value is equal to the peak value divided by the square root of 2.

Rotary Encoder : A form of photoconverter which is employed as an analog/digital convertor.

Rumble : It refers to undesired low-frequency noise heard on the loudspeaker of a gramophone. It is as a result of mechanical vibrations of the record adding to the signal produced from the pick-up cartridge.

S

Sampling : It refers to a technique in which only some portions of an electrical signal are measured and are used to produce a set of discrete values which gives the information contained in the whole.

A Sampling circuit is employed to generate a set of discrete values representatives as the instantaneous values of the input signal.

It also refers to a technique in which intermittent measurement of an electrical signal all made.

It is also refers to a system of quality control of mass-produced electronic components, circuits, devices, or other equipment.

Satellite : It means an artificial body which is projected from earth to orbit either the earth or another body of the solar system. There are two main classes *information satellities* and *communications satellites*.

Information satellites transmit signals from earth by means of highly directional aerials and return them to another earth location for purposes of long-distance telephony, TV broadcasting, etc.

Saturated Mode : It means the operation of a field-effect transistor in the portion of it s characteristic beyond pinch-off, i.e., Vp≤ where Vp is the pinch-off voltage and Vds the drain voltage.

It also refers to the maximum possible degree of magnetization of a material and is independent of the strength of the magnetic flux density applied to the material

Saturation Current : It means the current indicated on the portion of the static characteristic of an electronic device where it is substantially constant and independent of voltage.

Saturation Current : It means the current indicated on the portion of the static characteristic of an electronic device where it is substantially constant and independent of voltage.

Saturation Resistance : The resistance which takes place between the collector and emitter electrodes of a bipolar junction transistor for a specified value of base current when the device is started as a result of the external circuit.

Saturation Signal : A signal received by a radar receiver which has an amplitute larger than the dynamic range of the receiver.

Saturation Voltage : It means the residual voltage between the collector and emitter of a bipolar junction transistor for a specified value of base current when the device is saturated

see saturation due to the external circuit. It should be noted that it is smaller than the operating voltage under nonsaturated conditions.

Sawtooth Oscillator : A relaxation oscillator which generates a sawtooth waveform.

Sawtooth Pulse : A pulse having a geometrical shape similar to one complete period of a sawtooth waveform.

Sawtooth Waveform : A periodic waveform whose amplitude charges approximately linearly between to values the time taken in one direction the active interval being very much greater than the time taken in the other. The shorter period is termed the flyback.

S Band : It refers to band of microwave frequencies ranging from 1.55 to 5.20 gigahertz.

Scaler : It is a device mostly employed in counting circuits that produces an output pulse whenever a *disignated* number of input pulse have been received. The scaling factor is the number of input pulses per output pulse.

Scaling Tube : A gas discharge tube having several anodes and which is employed in a scaler circuit.

Scanner or Rectilinear Scanner : A device which visualizes the distribution of a radioactive compound in a particular system particularly the human body. The scanner is made of a scintillation crystal photomultiplier tube and amplifying circuits.

Double headed scanners have been developed in which the output is the sum of the outputs to two crystals which simultaneously scan the area of interest.

Scanning : Causing one complete horizontal or vertical traverse of the spot of light on the screen of a cathode ray tube (CRT).

It also refers to exploring a particular area or volume in a methodical manner in order to produce a variable electrical signal whose instantaneous values are a function of the information contained in the small area examined at each instant. The information scanned may then be reproduced by a suitable receiver. The technique is generally employed in television radar and facsimile transmission.

The most common type of scanning used in television and facsimile is rectilinear scanning

High velocity scanning is a type of electron scanning

Corse scanning is frequently used in order to produce

a rough picture of the target before carrying out a more detailed investigation.

Radar systems may also employ circular scanning in which one complete scan is a horizontal rotation of the radar beam through 360.

Generally scanning assumes that the scanning speed is uniform throughout each scan.

Scattering Loss : It refers to energy loss from a beam of electromagnetic radiation due to deflection of the radiation by scattering.

Schering Bridge : It refers to four-arm bridge for the measurement of capacitance (see diagram). At balance we have

$$C_x = C_a R_b / R_a$$

$$R_x = R_a C_b / C_s$$

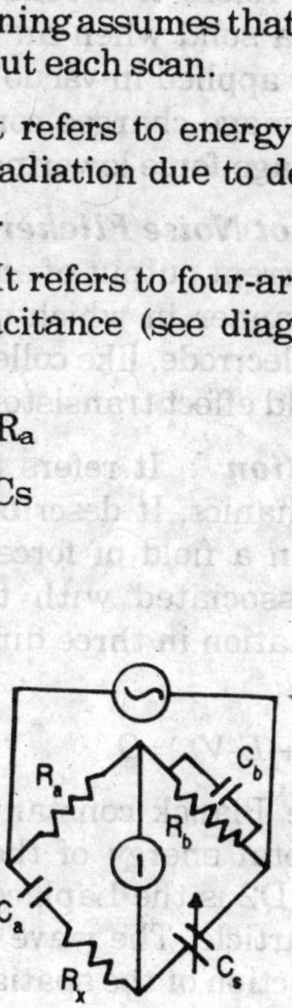

Schering bridge

Schmitt Trigger : A bistable circuit having output voltage level which is binary , is determined by the magnitude of the input signal, and is independent of the input signal waveform

The Schmitt trigger can be used in binary logic circuits in order to maintain the integrity of the logical one and zero levels.

Schottky diode or Hot Carrier Diode : A rectifying diode formed from a junction between a metal and a semiconductor a Schottky barrier (see Schottky effect). The term is employed for Schottky barriers in which the energy gap and doping level of the semiconductor are such that minority carriers in the semiconductor do not contribute significantly to the current flowing in the diode.

Schottky Effect : It refers to a reduction in the effective work function of a solid when an external accelerating electric field, F, is applied in vacuo to the surface. In the case of a metal, image charges contribute to the effect, which is called image-force lowering.

Schottky noise or Shot Noise Flicker Noise : The changes in the current output of an electronic device fue to the random manner in which electrons or holes are emitted from an electrode, like collector of a transistor or the source of a field effect transistor.

Schrodinger's Equation : It refers to fundamental equation of wave mechanics. It describes the behaviour of a particle moving in a field of force and considers to be broglie waves associated with the moving particle. Schrodinger's equation in three dimensions is expressed as

(h2/8n2m)v2^+(E-V)^=O

where h is the Planck constant, m the mass of the particle. E the total energy of the particle, and V the potential energy. D2 is the Laplace operator ^ the wave function of the particle. The wave function is a complex mathematical function of the spatial

Scintillation : It refers to a flash of light produced in certain materials (scintillators) when exposed to ionizing radiation

It means a rapid fluctuation of the image of the target of a radar system about its mean position on the radar display.

It also refers to a small random fluctuation of the received signal in a radio transmission system about the man signal value. This effect is similar to the twinking of light from star and is due to small variations in the density of the atmosphere.

Scintillation Counter : A radiation counter which is made of a scintillator (see scintillation), photomultiplier, amplifier, and scaler and is used to measure the activity of a radioactive source.

The scintillation counter is energy-dependent because the frequency of the emitted light is dependent on the energy of the incident radiation.

Scotophor : It refers to a material like an alkali halide, that can be used on the screen on a cathode ray tube instead of the usual phosphor when daylight viewing and long persistence are required. The material darkens under electron bimbardment to a black-on-white picture which may be erased by heating.

Scramble : A circuit or device which is employed in communication systems to produce an unintelligible version of the signal to be transmitte in a predetermined manner.

Screen : It refers to the surface of a cathode ray tube, suitably coated, on which the visible pattern is produced.

It also refers to shield which is a material which is suitably arranged in order to prevent or reduce the penetration of an electric or magnetic field into a particular region.

Scribing : It means scoring a wafer of semiconductor with a precision diamond tool in order to separate individual chips, each having a circuit or component, preparatory to packaging.

Scribing Channel : It means a gap left between the area on a semiconductor water during the manufacture of several circuits of components on the wafer, in order to allow for scribing into individual chips.

Second, Symbols : The SI unit of time regarded as the duration of 9.192631770 periods of the radiation corresponding to the transition between two hyperfine levels of the ground state of the caesium-133 atom. It was earlier defined as 1/86400 of the mean solar day

Secondary Electron : An electron emitted from a material due to secondary emission.

Secondary Emission : The emission of electrons from the surface of a metal, due to bombardment by high-velocity electrons or positive ions.

Secondary Standard : It refers to a copy of a primary standard which has a known difference from the primary standard.

It refers to a quantity which is accurately known in terms of a primary standard and may be used as a unit.

Standard Voltage : It refers to the voltage developed across the secondary (output) winding of a transformer.

It also refers to the voltage developed by a secondary cell.

Seed Crystal : A small single crystal which may act as a focus for crystallization, as from a supersaturated solution or a super-cooled liquid. It is employed to generate large single crystals of a semiconductor during the manufacture of solid-state electronic components, circuit, or devices.

Selective Interference : Interference to radiofrequency signals which take place only within a particular narrow band of signal frequencies.

Selectivity : The ability of a radio receiver to discriminate against radiowaves which have a carrier frequency different from that of which the receiver is tuned.

Selenium Symbol Se : A semiconductor element having atomic number 34. In the form of its grey allotrope selenium is extensively used in photononductive photocells.

Selenium Rectifier : A Schottky diode which consists of a selenium-iron junction and is used as a rectifier.

Self-Bias. It refers to required magnitude of bias developed from the main power supply at a point in a circuit by means of a dropping resistor.

Self-Capacitance : The inherent distributed capacitance associated with an inductance coil or resistor.

Self-Excited : It refers to an oscillator that produces a build up of output oscillations to a steady value upon application of power to the circuit.

Self-Inductance : For a circuit it is that property whereby any change in the current flowing in a circuit induces and emf opposing the charge and by virtue of which energy is stored in the magnetic field established by the current.

Self-Scaling Capacitor or Mansbridge Capacitor : A type of capacitor which is relatively little affected by puncture of the dielectric arising from an excessive voltage surge.

Semiconductor : Any material having electrical property between that of a conductor and an insulator is called a semiconductor. The conductivity of such a material depends upon atomic structure and certain influences such as light, heat or nearby electric fields.

The action of an external electric field results in conduction both in the conduction band and in the valence band. The conduction electrons are accelerated by the external field and carry negative charge through the semiconductor. The electrons in the valence band move to occupy each vacancy and the net effect is that a vacancy moves through the material as it it were a positive charge (Fig.a). The vacancies are termed as holes and are treated as positive charge carriers.

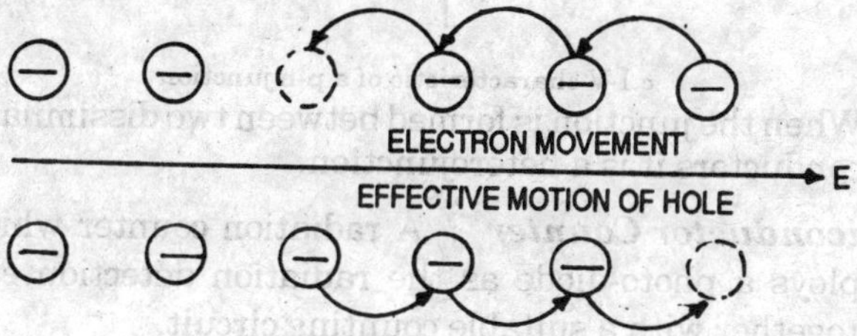

a. Hole conduction

When an electric field is applied to the semiconductor, charge carries move under the influence of the field but still undergo scattering processes. The result of the field is to impose a drift in one direction on to the random motion of the carriers (Fig.b). The average distance travelled by a minority carrier during its lifetime is called the diffusion length.

Junctions between semiconductors or between a metal and a semiconductor form a fundamental part of modern electronic components and circuits. A *p-n* junction is obtained between a *p*-type and an *n*-type semiconductor. A typical current-voltage characteristic is shown in Fig.c.

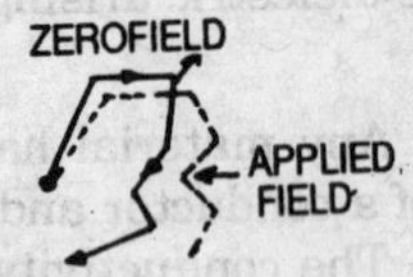

b. Movement of a carrier in a semiconductor

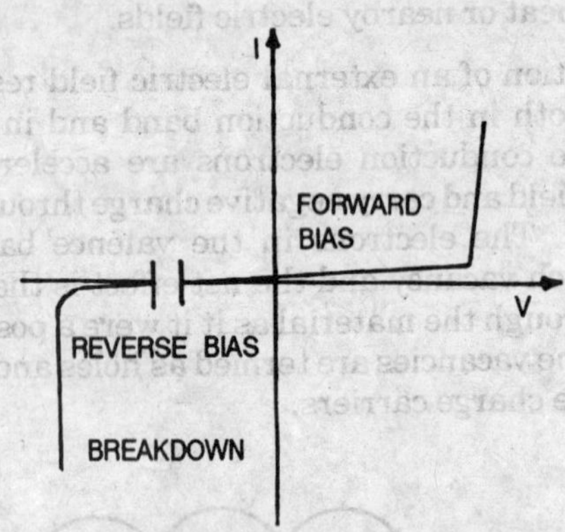

c. I-V characteristic of a p-n junction

When the junction is formed between two dissimilar semiconductors it is a heterojunction.

Semiconductor Counter : A radiation counter which employs a photo-diode as the radiation detection element together with a suitable counting circuit.

Semiconductor Device : Any electronic circuit or device which employs its operation on the flow of charge carries within a semiconductor.

Semiconductor Diode : A diode made from semiconducting material.

Semitransparent Cathode : A photocathode which shows photoemission from both surfaces in response to electromagnetic radiation incident on one of the surfaces.

Sensitivity : It refers to the change produced in the output of a physical device per unit change in the input.

Sequential Control : It means an operation of a computer in which a sequence of instructions is produced and input to the computer during the solution of a problem.

Serial Memory : Or sequential access memory. A memory employed in a digital computer having information retrieved by sequential access, i.e., in order to retrieve information stored as the *n*th record the preceding *(n-1)* records must first be scanned.

Serial transfer : It refers to transfer of information along a single path in a computer or data processing system in which the characters move one after the other along the path.

Serial Transmission : A communication system having characters of a word transmitted sequentially along a single line.

Series : Circuit elements are regarded in series if they are connected so that one current flows in turn through each of them. In case of *n* resistors in series the total resistance, R, is given by where *r1, r2..rn* are the values of the individual resistors. In other words resistors behave collectively as one large resistance.

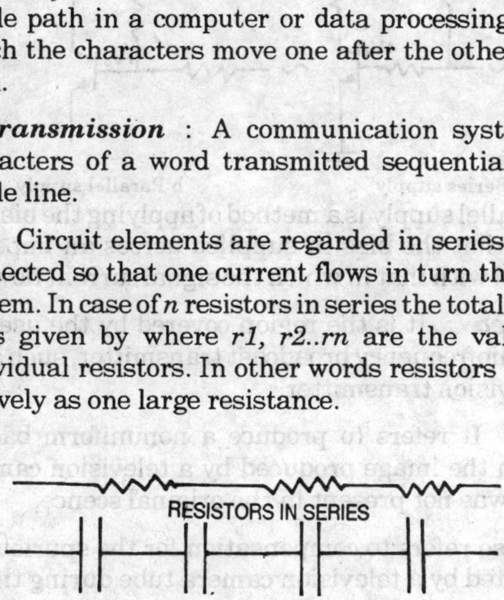

Capacitors in series

In case of *n* capacitors in series the total capacitance, *C* is given by

1/C=1/c1+1/c2+ 1/cn

where *c1, c2 cn* are the individual capacitances.

Series Parallel Connection : It refers to an arrangement whereby electronic devices or circuits may be connected either in series or in parallel.

It also refers to a method of connecting the elements of a circuit of network containing resistors, inductors, and capacitors so that some are in series and some in parallel with each other.

Series Supply : A method of applying bias to an electrode of an active device in such a way that bias is applied across the same impedance in which the signal current flows (Fig.a).

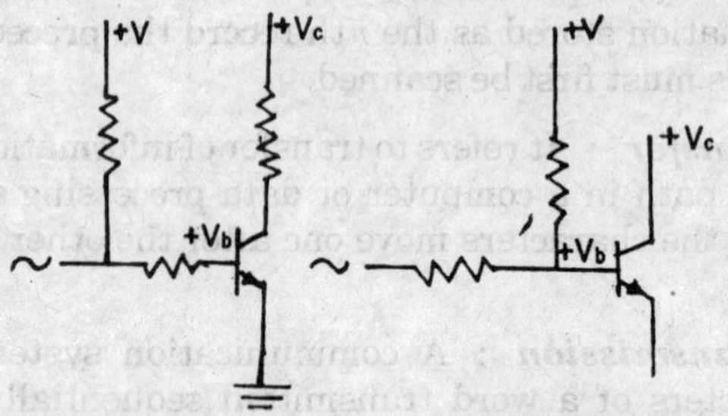

a Series supply b Parallel supply

Parallel supply is a method of applying the bias we such a way that the bias in supplied across an impedance in parallel with that in which the signal current flows (Fig.b).

Service Area : It is the region covered by the useful range of a radiofrequency broadcast transmitter, such as a radio or television transmitter.

Shading : It refers to produce a nonuniform background level in the image produced by a television camera tube which was not present in the original scence.

It also refers to compensation for the spurious signals generated by a television camera tube during the flyback interval.

Shadow Effect : It refers to an effct observed when electromagnetic waves are tansmitted, as during broadcast transmissions, due to the topograpaphy of the region between the transmitter and receiver

Shaped-beam Tube : A cathode-ray tube which is employed to display characters or numerals.

Shell-type Transformer : A transformer having the windings which are enclosed by the core. The core is made from laminations and mostly the windings are assembled and then the laminated core built up around them.

Shielded Pair : A transmission line which is made of two wires enclosed in a metal sheath.

Shift Register : A digital circuit which can store a set of information in the form of pulses and displace it either to the left or right upon application of a shift-pulse.

Shock Excitation : The production of free oscillations in a system, like a resonant circuit, by suddenly sending electric energy into the system; the energy may be in the form of a relatively short-lived impulse voltage or the shock may be the switching-on of a source of constant direct voltage.

Short Circuit : It refers to an accidental or deliberate electrical connection of relatively very low resistance between two points in a circuit.

Short-wave : A radiowave having a wavelength in the range 10 to 100 metres, i.e., in the high frequency band.

Short-wave Converter : A frequency changer in a radio receiver which is tuned to receive a broadcast transmission in the short-wave frequency band and converts it to a frequency band within the dynamic range of a standard receiver.

Shunt : A resistor, having a relatively low value, which is connected in parallel with a measuring instrument, like a galvanometer. Only a fraction of the current in the main circuit passes through the instrument in such a way that the shunt increases the range of the instrument and also protects it from possible damage caused by current surges.

Siemens : Symbol:S. The SI unit of electrical conductance, susceptance, and admittance. An element possesses a conductance of one siemens n case it has electrical resistance of one ohm. The siemens is now used in place of mho.

Siemen's Electrodynamometer : An electodynamometer which is calibrated as an ammeter, voltameter, or wattmeter.

Signal : A variable electrical parameter, like current or voltage, which is employed to convey information through an electronic circuit or system.

Signal Generator : Any electronic circuit or device which gives a variable and controllable electrical parameter.

Signal Level : It means the magnitude of a signal at a point in a transmission system, which reference to an arbitrarily chosen reference signal.

Signal to noise ration : The ratio of one parameter of a desired signal to the same or a corresponding parameter of the noise.

Silent Discharge : An electrical discharge which takes place at high voltages and is inaudible to the human ear.

Silent Zone : The portion of the skip area surrounding a particular transmitter which falls outside the range for ground-wave propagation of the transmitter.

Silica : Symbol SiO2. An extremely abundant compound present in several different natural forms, the best known of which are probable quartz and common sand (in which the silica is discoloured by ferric oxides).

Silica Gel : Deliquescent crystals consisting mainly of silica (SiO2) which are employed as a drying agent, particularly during dispatch and delivery of electronic and electrical equipment.

Silicon : Symbol Si. A semiconductor element having atomic number 14. It is found abundant in nature in the form of silicon dioxide (Silica) and is the most widely used semiconductor in solid-state electronics. It is cheap and extremely versatile and rapidly replaced germanium except for a very few specialized applications.

Silicon-controlled rectifier : (SCR) syn. reverse blocking triode thyristor. It refers to a ***pnpn*** device in which the forward anode-cathode current is controlled by means of a signal applied to a third electrode, known as the gate.

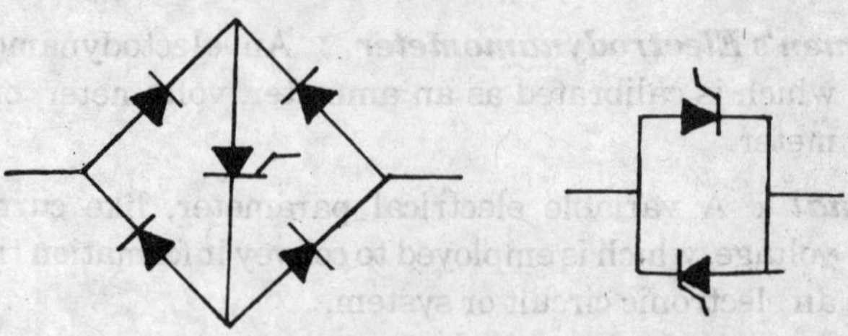

Full-wave rectification using an SCR

In case the gate turn-on current is supplied at the beginning of the positive half-cycle, a single SCR acts as a half-waverectification can be achieved by employing a bridge rectifier circuit in conjunction with an SCR or by using two SCRs in antiparallel connection.

Silicon-controlled Switch (SCS) : A semiconductor-controlled switch made in a silicon ship.

Silicon Diode : A semiconductor diode fabricated in silicon.

Silicon Rectifier : A silicon diode used as a rectifier.

Silicon Solar Cell : A solar cell constructed as alternate layers of *p - and n-* type silicon.

Silver : Symbol Ag. A metal, atomic number 47. The metal is very good conductor of electricity.

Simplex Operation : It refers to operation of a communications channel in one direction only.

Simulator : A device, which simulates the behaviour of an actual physical system and used to solve complex problems associated with the operation of the system.

Singing : It refers to undesired self-sustained oscillations occurring in a telecommunication system. The term is particularly applied to a telephone line that contains one or more repeaters. A musical note is generally obtained hence the term 'singing.

Singing arc : An arc which emits a musical note as a result of impressed oscillations that cause a variation in the hearing effect.

Single Crystal : It refers to crystal in which the corresponding atomic planes are effectively parallel.

Single-current System : A telegraph system which employs a uni-directional electric current for the transmission of the signal.

Single-ended : It refers to a process of supplying an input signal or obtaining an output signal from a circuit in which one side of the input or output is connected to earth.

Single-phase : It refers to an electrical system or apparatus which has only one alternating voltage.

Single-pole Switch : A swith which operates in only one cscuit Double-Pole switch.

Single-Short Trigger : A trigger circuit having each triggering pulse initiating only one complete cycle of events in a driven circuit, the cycle ending in a stable condition.

Single-sideband Transmission (SST). : The transmission of only one of the two side bands given by amplitude modulation of a carrier wave.

The important advantages of SST compared to transmission of the carrier and both side bands are the reduction in transmitter power required for the transmission and the reduced bandwidth required for the transmission of signals within a designated frequency band.

Single-tuned Circuit : A resonant circuit which may be indicated as a single capacitance and a single inductance with the associated values of resistance.

Sinusoidal : It refers to a periodic quantity which has a waveform graphically identical in shape to a since function.

SI Units : The internationally agreed system of units used for all scientific and technical purposes. The principle for the system MKS system and replaces the CGS and Imperial systems of units. There are three types of units in the SI system:*base units, derived units,* and *supplementary units.* All are absolute units.

The base units are an arbitrarily defined set of dimensionally independent physical quantities. If any purely mechanical system of units only three base units of mass, length, and time-are required. In a consistent electric and magnetic system four base units are required. In the SI system there are seven base units. These are the metere, kilogram, second, ampere, kelvin, candela (the standard of luminous intensity), and mole (the standard of amount of substance).

The derived units are formed by combining, by multiplication and/or division, two or more base units. Thus the coulomb, which is the derived unit of charge, is formed from a combination of one ampere times one second.

A set of 14 prefixes, including micro-and kilo, are used with the SI units to form multiples and submultiple of the units.

When considering electric and magnetic quantities a fourth term is required in addition to the fundamental units of mass length, and time, for their complete definition. In the SI system it is the ampere that is the fundamental unit.

Skew : It refers to time delay between any two electrical signals.

Skiatron : A type of cathode-ray tube which has a screen coated with a scotophor like potassium choloride, and thus gives a black-on-white picture.

Skin Effect : It refers to a nonuniform distribution of current over the cross section of a conductor when carrying alternating current, which the greater current density located at the surface (or 'skin') of the conductor.

Skip Distance : It is regarded as the minimum distance, for a specified operating frequency, at which radiowaves are received in a given direction from the transmitter by reflection from the ionosphere.

Slave Circuit : An electronic circuit which requires a trigger or a clock pulse from an external source in order to perform its cycle of operation.

Slide Wire : A wire of uniform resistance having a sliding contact which can make a connection at any desired point along the length. Slide wires are used to provide two variable resistance, as a potentiometer, or to provide a variable resistance ration. An overall length of one metre is mostly chosen.

Slope Resistance : It refers to the ratio of an incremental voltage change applied to a specified electrode in and electronic device to the corresponding current increment of the electrode, the voltage applied to all other electrodes being maintained constant at know magnitudes.

Slow-bleak Switch : It refers to a manually operated switch whose speed of operation depends upon the speed at which the operating handle or lever is moved.

Smoothing Circuit : A circuit which is designed to reduce the amount of ripple present in an essentially unidirectional current

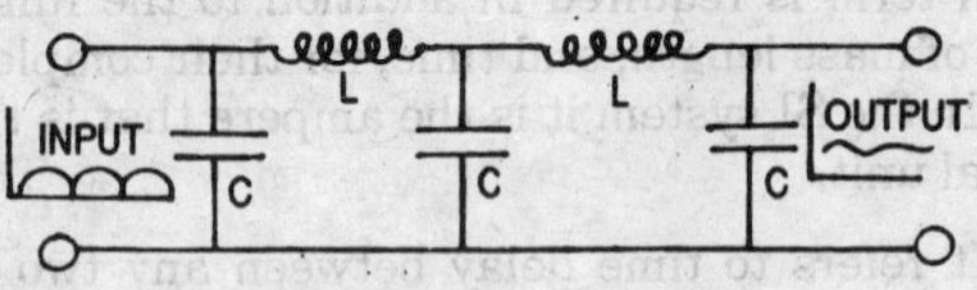

Simple low-poss filter as smoothing circuit

or voltage. A typical smoothing circuit consists of a low-pass filter (see diagram) but a single inductance may be employed.

Snow : An undesired pattern appearing on the screen of a television or radar receiver which resembles falling snow.

Solar Cell : A device which employs the photovoltaic effect in order to convert radiation from the sun directly into electrical energy. The solar cell is essentially a *p-n* junction that has a relatively large surface area and relatively high efficiency (of the order of 10% to 15%). A typical equivalent circuit is shown (Fig.a).

Solar cells are the most important long-duration power supply for satellites and space vehicles. They have been made from a range of different semiconductors including silicon, gallium arsenide, selenium-cadmium sulphide, and thin-film cadmium sulphide. Gallium arsenide provides the optimum inherent efficiency at lower temperatures (from 0~C). At higher temperatures (above 100°C) a 50-50 gallium arsenide/gallium phosphide material produces the optimum.

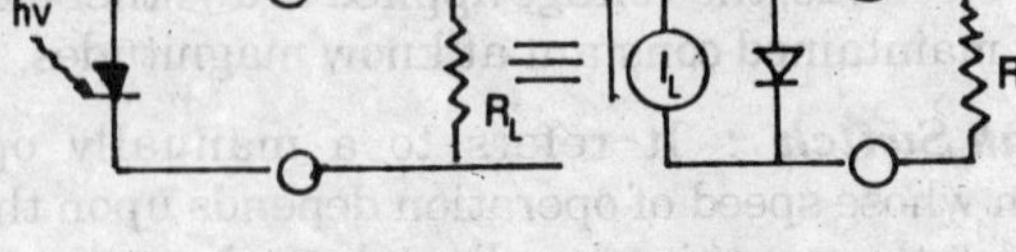

b Equivalent circuit of an idealized solar cell.

Solenoid : An inductor with a movable iron core is called a solenoid. These are used to close switches and valves.

Solid Conductor : A conductor which is composed of a single wire or uniform thin metal rather than being standard or otherwise devided.

Solid-State Camera : A television camera containing the light-sensitive target area which consists of an array of charge-coupled devices (CCDs). Exposure to light energy gives the generation of electron-hole pairs in the semiconductor substrate, the number of electron-hole pairs generated is a function of the light intensity.

Solid-state Device : It refers to an electronic component or device which consists of solid materials, that depends for its operation on the movement of charge carries within it. A solid state device has no moving parts.

Solid-state Memory : A memory formed as an integrated circuit in a chip of semiconductor.

Solid-state Physics : The branch of physics which deals with the structure and properties of solids and any associated phenomena.

Sonar : A process of detecting and locating underwater objects which operates on a similar principle to radar but the transmitted pulse is a burst of sound energy, generally ultrasonic, rather than radiofrequency waves.

A sonar system which is employed to measure the depth of the sea bed by projecting the pulse vertically downwards is termed an *echo-sounder*. Any system which discovers submarines is termed asdic. One type of asdic employs *echo-ranging*.

Source : The device which produces electrical energy, is called a source.

It also refers to the electrode in a field-effect transistor which charge carriers (holes or electrons) to the inter-electrode space.

Source Impedance : The impedance presented to the input terminals of an electronic circuit or device by any source of electrical energy.

Space Charge : The number of electrons emitted from the cathode depends on the operating temperature of the cathode according to Richardson's equation. The potential applied to the plate does not affect the number of electrons emitted from the cathode. However the number of electrons reaching the plate will depend on the potential difference between the plate and the cathode. The accumulation of electrons between the plate and the cathode is called space charge.

Space-Charge Density : It means the net electric charge per unit volume in a space-charge region.

Space Wave : A radiowave which travels between a transmitting and a receiving aerial situated above the ground that includes the direct wave and the ground-reflected wave.

Spark : It refers to visible disruptive discharge of electricity between two points of high potential difference, preceded by ionization of the path.

Spark Channel : It refers to the path followed by a spark between two electrodes.

Spark Counter : A radiation detector which is employed to detect and count alpha particles.

The alpha particles may be detected by ear due to the noise associate with the sparks. They may be counted by using a suitable counting circuit which detects the voltage drops across the anode load resister or bay photography.

Alpha counters are completely unaffected by beta particles and X-or gamma rays and are thus very useful for the detection of leakage of alpha-emitter (such as radium) from sealed containers that normally completely absorb the alpha particles.

Spark Photography : Any form of photography which is employed as a spark discharge as the source of illumination.

Speaking Arc : A means of reproducing sounds which uses an arc discharge to amplify the original sound.

Spectral Characteristic : A graph which indicates the sensitivity or relative output as a function of frequency for any frequency-dependent device, circuit, or other equipment.

Speech Recognition Device : A device which is obtained from the human voice coded signals for a given computer.

Spin : The intrinsic angular momentum possessed by all elementary particles, including the electron and proton.

The magnetic fields obtained by an atomic electron spin an orbital motion interact in such a way that the electron can exist in either of two closely spaced energy levels. The spin in one level is +1/2 and in the other it is 1/2. Due to its spin and electron has an intrinsic magnetic moment.

Spot : It refers to an area, like on the screen of a cathode-ray tube, which is immediately affected by an electron beam impinging on or near to it.

It also refers to a local imperfection on the surface of an electrode.

Spot Speech : It means the product of the number of spots into the scanning line and the number of scanning lines per second in a television picture tube.

The number of spots scanned or recorded per second in facsimile transmission.

Sprat : A hand-held radar system which is based on the Gunn diode. It is about 2.5 kg and has a range of about 600 meters.

Spreading Resistance : It means the component of the series resistance of a semiconductor device which is because of the bulk material located some way from the junctions and metal contact.

Spurious MOST : An undesired MOS field-effect transistor formed in an MOS integrated circuit by the interconnections between the parts of the circuit.

Spurious Response : It refers to an output from an electronic circuit, device, or transducer in the absence of an input signal or due to the presence of an input signal other than the required signal.

Sputtering : The use of a high-voltage low-pressure glow discharge deposit material in the form of a film.

Square-low Detector : A detector which generates an output voltage which is proportional to the input voltage.

Square-Loop Material : A ferromagnetic material which

shows a substantially rectangular magnetic hysteresis curve. The ration of the remanence to the saturation flux density of this type of material is the *squareness ratio*.

Square Wave : A pulse train which has rectangular pulse the markspace ration of which is unit (see diagram).

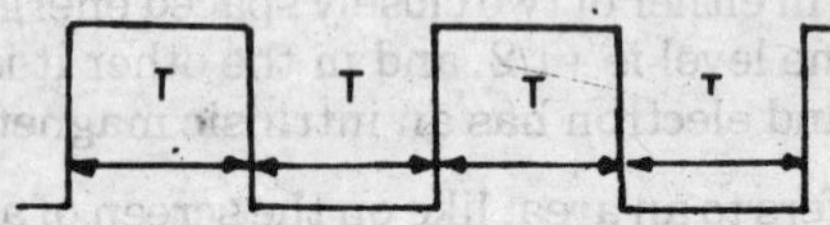

Square wave

Square-Wave Response : The response of any electronic circuit or device to a square-wave input signal, given as the wave shape and peak-to-peak amplitude of the output signal is know as square-wave response.

Squegging Oscillator : The oscillator in which the main oscillations output from it change the electrical conditions of the oscillator circuit in such a way the output amplitude periodically builds up to a peak value and then falls to zero.

Squelch Circuit : A circuit employed in a radio receiver which suppresses the audiofrequency output unless an input signal of a predetermined character is received.

Stabilivolt : A gas tube, like a neon lamp, which generates an esentially constant voltage drop across its terminals for a relatively wide range of currents.

Stabilization : The provision of negative feedback in a circuit, like an amplifier, which has inductance and capacitance and introduces substantial gain between the input and the output, in order to provide overdamping (see damped) in the circuit and thereby prevent any oscillations occurring.

Stable Circuit : A circuit that does not produce any undesired oscillations over the entire dynamic operating range.

Stable Oscillations : Oscillations which tend to decrease in amplitude with respect to time.

Standard Cell : An electrolytic cell which is employed as a voltage reference standard.

Standardization : It is the method of relating a physical magnitude or the indicated value of a measuring instrument, such as a volt meter, to the primary standard unit of that quantity.

It also refers to a nationally or internationally agreed system for the standardization of electronic or electrical components or devices.

Standarding Wave or Stationary Wave : A wave which remains stationary, in other words the displacement at any given point is always the same and a given displacement, such as that of a nodem is not propagated along the wave. These waves are due to the superimposition of two or more waves of the same period and usually occur when a wave is reflected totally or partially from a given barrier.

Starter Electrode : An auxiliary electrode which is employed in a glow-discharge tube in order to initiate the glow discharge; it is also used in an arc discharge tube, such as a mercury-arc rectifier, to initiate the arc discharge.

Starting Current : It refers to the magnitude of the current in an oscillator at which self-sustaining oscillations are initiated for specified load conditions.

Start-Stop Apparatus : An instrument employed in telegraphy in which each set of coded signal elements corresponding to a transmitted character is preceded by a start signal that activates the receiving apparatus in preparation for the character and is followed by a stop signal which brings the receiving apparatus to the rest. Start-stop apparatus finds applications in telex systems.

Stat : A prefix to a unit, indicating it s use in the CGS electrostatic system of units.

1statampere=3.336x10-10 ampere

1statvolt=2998x102 volts

1 statohm=8.981x1011 ohms

Static : It refers to undesired random noises generated at the loudspeaker of a radio receiver.

Static Memory : A solid-state memory which does not require refresh operations in order to retain the stored information.

Stationary State : The state of an atom or other system which in the quantum theory or quantum mechanics is described by a given set of quantum numbers. Each of the various energy states which may be assumed by an atom is a stationary state of that atom.

Steady State : It refers to a state reached by a system under steady operating conditions after any transient effects resulting from a change in the operating conditions have died away.

Steerable Aerial : A direction aerial which is made of several fixed units in which the direction of a maximum sensitivity can be altered by adjusting the phase relationship between the units from which it is formed.

Stepping Relay : A relay with a contact arm which rotates through 360 in two or more discrete steps.

Step Stress Life Test : A life test having several stress levels applied sequentially for periods of equal duration to one sample.

Storage Device : Any device which keeps information by a physical or chemical method. The term is mostly used to a memory in a computer, which holds information usually in binary form.

Storage Time or Retention Time : In any device which stores information, such as a storage device or a storage tube, maximum time that the information may be stored without significant loss of information.

Storage Tube : An electron tube which is employed to store information for a determined and controllable time.

Strapping : It refers to a process of suppressing undesired modes of oscillation of multicavity magnetron by connecting together resonant segments of the same polarity.

Stray Capacitance : Any capacitance in an electronic circuit or device which is as a result of interconnections, electrodes, or the proximity of elements in the circuit and is additional to the intentional capacitance of the circuit or device.

Striking Potential : The potential difference which is required to initiate an arc discharge between two electrodes.

String Electrometer : An electrometer which consists of a fine metallized quartz fibre stretched between to parallel conducting plates. The plates are oppositely charged by the potential difference to be measured and the deflection of the quartz fibre observed.

Strip Core : A magnetic core which is constructed from a continuous strip of the magnetic material.

Stroke Pulse : A pulse employed in a digital computer or radar system for timing or indicating purposes.

Strowger Exchange : An automatic telephone exchange having the necessary connections between telephone lines and exchanges are effected by a series of Strowger switches in order to establish a telephone channel.

Stub : A device employed for impedance matching of a transmission line to a load in microwave and ultra high frequency applications.

Subassembly : Two or more parts which form a portion of an assembly but that may be individually replaceable.

Subcarrier : A carrier wave which is employed to modulate a second different carrier.

Subsonic Frequency : It is regarded as a frequency of value less than the audio frequency range.

Substandard : It means a measuring instrument which is employed as a standard instrument but is not quite so accurate as the primary standard.

Substation : A complete assemblage of plate, equipment and the necessary buildings at a place where electrical energy is received and where it may be either converted from alternating current to direct current, stepped up or stepped down by means of transformers, or used for control purposes. The substation mostly receives power from one or more power stations.

Substrate : It refers to single body of material on or in which circuit elements or integrated circuits are fabricated.

Sub system : It refers to interconnected set of related circuits that form a logical sub division of a piece of electrical equipment or an operational system and may be manufactured as an assembly or subassembly.

Summation Instrument : A single instrument which receives signals, like current, power, or energy from a number of separate circuits and measures the agtregate.

Superconductivity : A phenomenon which takes place in certain metals and a large number of compounds and alloys when colled to a temperature close to the absolute zero of thermodynamic temperature.

superconductor : A material which shows superconductivity.

Superheterodyne Receiver : This type of receiver is commonly used now a days. The predetactor amplification is done at a constant intermediate frequency irrespective of the frequency of the carrier waves of the transmitter whose signal is to be received.

Supermalloy : *Tradename*. It is an alloy of iron, nickle, and molybdenum which is similar o permalloy but hat has a higher magnetic permeability.

Supermendur : *Tradename*. A square-loop material employed as a magnetic core.

Super Regenerative Reception : A process of reception used for ultra high-frequency radiowaves in which the detector is a squegging oscillator.

Supersensitive Relay : An electromechanical relay which operates with currents of less than about 250 microamps

Supervisory Control : A remote-control system which employs relatively few transmission channels.

Suppressed-Carrier Transmission : A process of transmission of radiowaves in which the carrier component of the modulated wave is not transmitted. One or both or the side bands only are transmitted.

Suppressed-Zero Instrument : A measuring or recording instrument having the zero position which falls outside the dynami range of the instrument.

Surface-Barrier Transistor : A transistor having the usual p-n junctions replaced by Schottky barriers.

Surface Conductivity : It is the reciprocal of surface resistvity.

Surface Resistivity : It means the resistance between two opposite sides of a unit square of the surface of a material. The measured value can change greatly depending on the method of measurement.

Surface Wave : A radiowave which moves along the surface separating the transmitting and receiving aerials.

Surge : It refers to an abnormal transient electrical disturbance in a conductor.

Surge-Current Indicator or Magnetic Link : A device which determines the approximate magnitude of a surge current.

Surge Diverter A device that is put between a conductor and earth and diverts the major part of any excessively large voltage surge. A *lightening arrester* is specific example of a surge diverter.

Surge Impedance : The impedance passed by a transmission line to an electrical surge which travels along the line as travelling wave.

Surge Modifier : A device connected to a power transmission system or telecommunication system which changes the waveshape of any surges that might occur in order to protect that equipment, circuits, or devices, associated with the system.

Susceptance : Symbol:B;unit;siemens. The *imaginary* part of the admittance, Y, which is given by the expression

$$Y=G+iB$$

Where G is the conductance. For a circuit having both resistance, R, and Resistance X, the suseptacnce is given by

$$B=X/(R2+X2)$$

Suceptibility.

(i) Magnetic susceptibility. Symbol:xm. It is a dimensionless quantity given by the equation

$$Xm=pr-1$$

where G is the relative permeability of a material. Magnetic susceptibility describes the response of a material to a magnetic field, being the ration of agnetization to magnetic field strength:

xm=M/H

X is a tensor when M is not parallel to H, otherwise it is a simple number. It has a wide range of values : diamagnetic materials have a negative value; ferromagnetic materials can have a very large variable value (up to about one), which is dependent on the magnetic field strength.

(ii)Electric susceptibility. Symbol:Xe. It is a dimensionless quantity given by the equation

Xe=er1

where er is the relative permittivity. Electric susceptibility measures the ease of polarization of dielectric and is given by the ratio

xe=P/eoe

where *p* is the dielectric polarization. *E* the electric field strength and *e O* the permittivity of free space.

Sweep Voltage : It is the voltage output from an internal or external time base that when applied to the appropriate deflector plates or coila of a cathode-ray tube result in a horizontal or vertical deflection of the electron beam.

Swing : The limits of the values of a varying electrical parameter, like amplitude or frequency.

Swiss Cheese : A technique of manufacturing circuits which employ a thin board containing a lattice of holes into which circuit elements, suitability encapsulated to form pellet shapes are inserted and attached.

Switch :

(1)A device which opens or closes circuit

(2)A device which results in the operation condition of a circuit to change between discrete specified levels.

(3)A device which selects from two or more components, part, or circuits the desired element for a particular mode of operation.

In general, a switch may consist of a mechanical device, like a circuit-breaker, or a solid-state device, like a transistor, Schottky diode, or field-effect transistor.

Switching Tube : An arc-discharge tube, like an ignitor, which has at least two anodes and is used a switch.

Synchrocyclotron or frequency-Modulated Cyclotron : A form of cyclotron that takes account of the relativistic mass change of the accelerated particles.

Synchronism : The relationship between two periodically changing quantities when they are in phase.

Synchronometer : An instrument which counts the number of cycles of a periodically varying quantity that occur during a perdeterminded time interval.

Synchronous Alterating-Current Generator or Alternator Synchronous Generator : An alternating-current generator which is made of one or more coils that are made to rotate in the magnetic fields produced by several electromagnets by direct current source.

Synchronous Clock : A mains-operated electric clock having the speed of the driving motor as a function of the mains frequency and therefore the time-keeping is controlled by mains supply.

Synchronous Computer : A computer having the timing of all operations controlled by a clock.

Synchronous Gate or Clocked Gate : A gate the output of which is synchronized to the input signal.

Synchronous Logic : A logic system which operates with *Synchronous timing*, i.e the timing of all the switching operations is controlled by clock puses.

Synchrotron : A cyclic particle accelerator which is employed to accelerated a beam of electrons electron synchrotron-or protons protons-proton Synchrotron-to very high energies.

The particles are mostly injected at high energies from a linear accelerator in order to minimize the range of modulation of the r.f filled a t lower energies, acceleration is then because of the electromagnetic induction only. At present proton synchrotrons produce the highest energies of accelerated particles.

T

Tandem : It is a method of connecting two quadripole network in such a way that the two output terminals of one network are connectted to the input terminals of the other.

Tangent Galvanometer : It is a moving-magnet galvanometer having a small magnet in the centre of a large fixed coil. The magnet carries a long light non magnetic pointer, gererally alumliniu, which moves over the scale.

The sine galvanometer is similar to the tangent galvanometer except that the coil and scale are rotated together in such a way that the neele is at zero.

Tantalum : Symbol Ta is a metal, atomic number 73, has an extremely high resistance to corrosion.

Tap Changer : Syn.ratio adjuster. It is a device which changes the ratio of a transformer by selecting the desired tapping. These are of two types i.e., 'On load tap changer' and 'off circuit tap changer'.

Tapping : It is a conductor, which makes an electrical connection with a point between the ends of a winding or coil.

Target : It is usually or designated object deltected by a radar or sonar system.

Target Voltage : It is the potential difference between the cathode and the signal electrode of a low-electron-velocity camera tube.

Tearing : It is an effect observed in the picture on the screen of a television receiver in which the picture appears to break up.

***Teflon*.** : Tradename Polytetrafluoroethylene (PTFE). It is an insulator which has an extremely high resistivity and is very resistant to moisture and temperature.

Telecomnunications : The study and means of the tr4ransfer of information by any electromagnetic means, like wire or radiowaves.

Telecommunication System : It is the complete assembly of apparatus and circuits required to effect a desired transfer of information. Systems include television, radio and telephony.

Telegraphy : It is communication using telecommunication system. The transmission is done by a suitable signal code, such as international Morse code, either by means of wire or by radio.

Telemetry : The measurement at a distance.

Telephony : It is communication by means of a telecommunication system which is designed to transmit speech or sometimes sounds.

Teleprinter : It is form of start tstop typewriter machine used for transmission receiption of typed information. The keyboard converts typed information into electrical signals for transmission and at receiver the process is reverse. They are generally used in telex-system.

Teletext : It is an information service by which information can be displayed as pages of text and pictorial meterial on the screen of a commercial television receiver.

Television : A telecommunication system in which both visual and audio information is transmitted for reproduction at a receiver. A television receiver detects the signals and gives an image on the screen of a specially designed cathode-ray tube and a simultaneous sound output from a 'loudspeaker'.

The basic television system transmits images in black and white only (monochrome television). Colour television is now widely employed and the broadcast signal is received on special colour receivers.

The transmitted signal contains both video and audio information.

Television Camera : It is a device employed in a television system to convert the optical images from a lens into electrical signals. The optical image formed by the lens system of the camera falls on to a photosensitive target. This is scanned, mostly by a low-velocity electron beam and the resulting output is moudulated with video information obtained from the target area.

The cmera consists of three major parts boused in one container. L optical lens system, camera tube, and preamplifier.

The camera employed in colour television has three camera tubes each of which receives information which has been selectivity filtered to provide it with light from a different portion of the visible spectrum. The original multicoloured signal breaks into red green,and blue components. The scanning systems in the three tubes are driven simultaneously by a master oscillator so that the output of each tube corresponds to the same image point. The three outputs are then combined to give brightness (luminance) and colour (chrominance) information.

Television Receiver : It is a device which receives a television broadcast signal and effects the sound and vision reproduction of the original scene.

Telewriter : It is device employed in telegraphy that converts manually controlled movements of a pen over a plane surface into two currents whose instantaneous values are a function of the position of the pen.

Telex : It is regarded as telegraphy system which helps people to communicatte directly and temporarily among themselves using teleprinters and the public telegraph system.

Temperature Coefficient of Resistance : This term is defined as the incremental change in the resistance of any material as a result of change in thermo-dynamic temperature.

Tera : It is a prefix to a unit, denoting a multiple of 10 12 of that unit.

Terminal : (a)) An instrument which provides input/output facilities to a computer, often from a remote location.

(b) Any point at which interconnecting leads may be attached to an electronic circuit or device and at which signals may be input or output.

Termination : A load impedance connected across the output of a transmission line or transducer which completes the circuit while ensuring impedance matching and preventing unwanted relections.

Tertiary Winding : It is an additional secondary winding on a transformer. It may be used to supply a load when a different voltage is required from that of the main secondary or when a load must be kept electrically insulated from that of the normal secondary.

Tesla : This is the SI unit of magnetic flux density. One tesla is defined as one weber of magnetic flux per square metre of circuit area.

Tesla Coil : It is defined as an induction coil used to generate very high frequency high voltage oscillatory currents. The induction coil, I discharges across a spark gap and feeds the primary, P, of a transformer through two large capacitors.(See diagram).

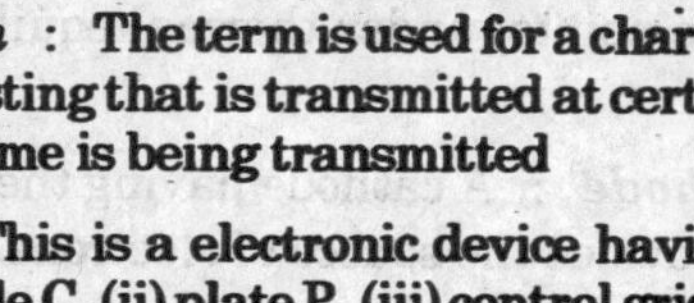

Tesla coil

Test Pattern : The term is used for a chart used in television broadcasting that is transmitted at certain times when no programme is being transmitted

Tetrode : This is a electronic device having four electrodes (i) cathode C, (ii) plate P, (iii) control grid S^I and (iv) screen grid S^2 (see Fig.). The control grid serves the same purpose as a single grid of a triode. The screen grid S^2 is intended to prevent oscillation in the circuits. It reduces the grid to plate capacitance and effectively minimises the feedbackaction. This helpls to reduce the instability in RF

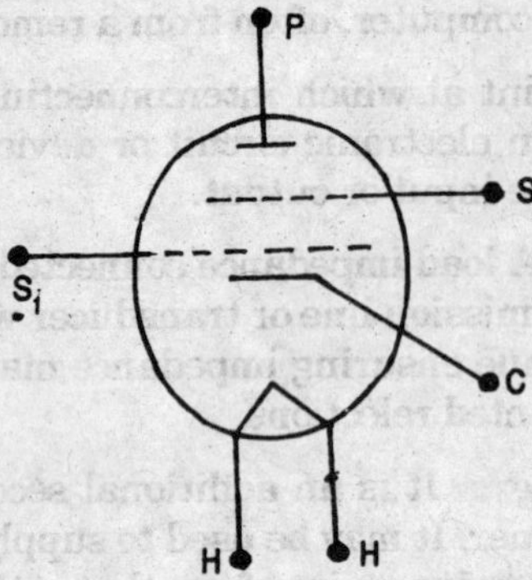

Tetrode

amplifiers. It the control grid S^2 is absent, then the voltages induced upon the control grid can be sufficiently great to start and maintain oscillations throughout the tube circuits.

Thermal Breakdown : Breakdown of a reverse-biased *p-n* junction due to generation of excess free charge carriers as a result of cumlative interation between incresing junction temperature and increasing power dissipation.

Thermal Imaging : It is getting an image of an object by means of the infrared radiation emited by it. A camera tube with a suitable lens system may be employed to produce the image.

Thermal Instrument : A measuring instrument which employs the hearing effect of a current in a conductor.

Thermal Resistance : The ratio of the difference in temperature between two specified points to the heat flow between these points, under thermal equilibrium conditions.

Thermionic Cathode : A cathode having thermionic emission which provides the source of electrons.

Therionic Emission : The electrom emission from the surface of a solid which is a result of the temperature of the material. An electron can escape from the surface with zero kinetic energy when it has thermal energy just equal to the work function of the material.

Thermionic Valve : It is defined as a nulti electrode evacuated electtron tube which has a thermionic cathode as the source of electtrons.

It should be noted that under conditions of reverse bias no current flows in the valve until the field across the valve is suffcient to cause field emission from the anode or arc formation; breakdown of the device will then occur.

Anode and ttransfer characteristics of a triode are shown in Fig.(a) and (b).

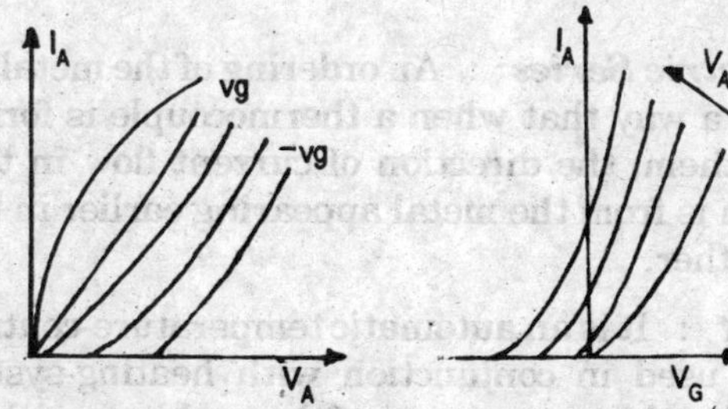

(a) Anode characteristics of a triode (b) Transfer characteristics of a triode.

Thermistor : It is defined as a resistor fabricated from semiconductor material which has a lage nonlinear negative temperature coefficient of resistance.

Thermoammeter : This is a type of ammeter which measures a current by means of the heating effect produced by the current. There are two types of thermoammeter, i.e., thermocouple-ammeter and hot wire ammeter.

Thermocouple : The term is used for two dissimilar metals joined at each end to form an electrical circuit. When the two junctions are maintained at different temperatures an electromotive force (e.m.f.) is developed between them. The e.m.f. is not affected by the presence of other metal junctions provided that they are all maintained at the same temperature. The thermocouple can be used with a suitable measuring instrument and used as a thermometer.

Thermocouple instruments are measuring instruments whih employ a thermocouple to measure electtric current, voltage, or power by means of the heating effect in a metallic strip or wire. A vacuum thermocouple is thermoammeter which is designed to meaasure very small currents.

Thermodynamic Temperature : It is syn. to absolute temperature. Symbol: T is defined as the temperature which is determined as a function of the energy possessed by the matter and as such is a physical quality that can be expressed in units, termed kelvin.

Thermoelectric Effects : The phenomenon which takes place as a result of temperature differences in an electrical circuit.

Thermoelectric Series : An ordering of the metal elements in such a way that when a thermocouple is formed from two of them, the direction of current flow in the hotter junction is from the metal appearing earlier in the series to the other.

Thermostat : It is an automatic temperature-control switch that is used in conjunction with heating-systems and maintains the temperature of the meddium within known limits.

Thevenin's Theorem : A theoren used to simplify the analysis of resistance networks theorem. According to this theorem if two terminals (A, B) emerge from a network in order to connect to an external circuit, thgen as far as the external circuit is concerned the network behaves as a voltage generator. The e.m.f. of the voltage generator is equal to $V_{A,B}$, where Va, D is measured under open-circuit conditions, and it has an internal resistance represented by

$$RA,B = Va,B / Ia,B$$

where *IA,B* is the short-circuit current.

Thick-Film Circuit : It is defined as a circuit that is manufactured by the thick-film techniques, like silk screen printing, and generally has only passive components and interconnections.

Thin-Film Circuit : It is a circuit, manufactured by thin-film techniques, such as cathode sputtering or vacuum evaporation, and generally has only passive components and interconnections.

Thin-Film Memory : It is a magnetic memory which is made by using thin-film techinques, like cathode sputtering or vacuum evaporation.

Thin-Film Transistor (TFT) : An indulated-gate field-effect transistor (IGFET) fabricated by using thin-film techniques on an insulating substrate rather than on a semiconductor chip.

The technique is now used mainly for the construction of silicon-on-sapphire C/MOS circuits.

Thoriated Tungsten Filament : A tungsten filament which has a small amount of thorium and is employed as a thermionic cathode which is called thoriated tungsten filament. The amouunt of thermionic emission of electrons at any given temperature is greater for the mixture than for pure tungsten.

Thou : One thousandth of an inch.

Three-Phase to Two-Phase Transformer : It is a transformer which is constructed to operate with a three-phase supply in the primary circuit and produce a two-phase supply at the secondary circuit, or *vice versa*.

Three-Phase Transformer : It is a transformer which has three independent sets of windings, each generally with the same turns ratio, and is suitable for use with a three-phase mains input.

Thresold Frequency : It is the frequency at which a particular phenomenon, such as the photoelectric effect or photoconductivity, just takes place or the frequency below which an electronic device, like a high-pass filter does not operate.

Threshold Voltage : It is defined as the voltage at which an electronic device first functions in a specific manner.

Thyratron : The term is used for a gas-filled tube having three electrodes in which the voltage applied to one of the electrodes-the grid (or control) electrode-is employed to initiate the discharge but does not limit it. The anode characteristic is shown in Fig.a.

It should be noted that the value of the grid voltage at which the tube will strike is a function of the applied anode

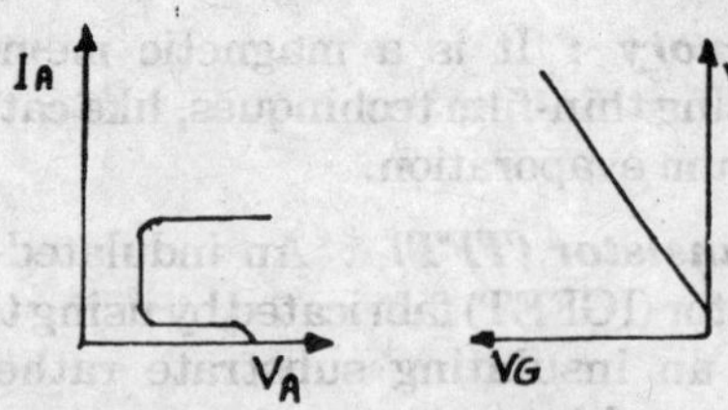

a Anode Characteristic. b Control characteristic

voltage and the device geometry. The control characteristic is the curve showing anode voltages and the corresponding critical grid voltage for striking a typical curve is shown in Fig.b.

Thyrister : The terms is used for a semiconductor device having three or more junctions and that has current-voltage characteristics similar to those of the thyratron.

Tight-Coupled Inductor : It is a mutual inductance which has very efficient coupling between the two self-inductances such that the square of the mutual inductnce is nearly equal to the product of the individual self-inductances.

Time base : Voltage which is a predetermined function of time that is used to feflect the electron beam of a cathode-ray tube that the luminous sopt traverses the screenin a desired manner.

The most common type of time base in one that produces a linear sweep.

Time Constant : It is regarded as the time required for a unidirectional electric quantity, such as voltage or current, to decrease to l/e of its initial value or to increase to (l - l/e) of its final value in response to a change in the electrical conditions in an electronic circuit or device. In other words at any instant following initiation of the change in electrical equilibrium, such as switching a d.c. supply voltage or off, instantaneous change of a quantity, V, is given by the equation:

$dV/dt=(V_f-V)/\tau$

where *V*, is the final value of the quantity and *t* the time contant. In the case of a decreasing quantity Vf is mostly zero and the expression can be written as

$dV/dt = V/\tau$

The time constant is a measure of the speed of operation of any circuit or device; circuits which contain capacitance or inductance can have very long time constants.

Time-Delay Switch : It is a switch which automatically causes a predetermined series of operations to be carried out which a specified time interval between each operation.

Time Discriminator : A circuit having the magnitude of the output as a function of the time interval between two input pulses and the sense of the output is determined by the ordr in which the inputs are received.

Time-Division Multiplexing : It is a type of multiplex operation that samples and transmits each of the input signals sequentially in such a way that each input signal is transmitted over the common channer during a set of predetermined time intervals.

Time Lag or Time Delay : The time which elapses between operation of a circuit-break, relay, or similar apparatus and the response of the current in the main circuit.

Time Sharing : It is regarded as a form of time-division multiplexing by which a numner of users may communicate directly with a computer by means of a numner of individual terminals.

Time Switch : It is a device which incorporates a clock mechanism and automatically operates at predetermined times.

Time to Trip or Opening Time : This term is used for natural delay, apart from any intentional delay, between the instant at which a predetermined signal is applied to a switch, circuit-breaker, or other similar device and the instant at which the device operates.

Tin : Symbol:Sn. It is metal, atomic number 50, which has good resistance to corrosion and alloys readily with other metals, such as a copper.

T-Junction : It is syn. to hybrid T-junction: magic-T junction.

Torque : It is defined as the moment exerted by a force acting on a body and tending to cause rotation about an exis.

Total Capacitancce : The term is designated for the capacitance between one conductor in a system and all the other conductors in the system when electrically connected together.

Total Emission : The peak value of the current which may be obtained by thermionic emission from a cathode under normal heating conditions.

Touch Control : A menually operated whitch which is activated by touch and that presents no moving parts to the operator.

Trace : It is the image traced out by the luminous sopt on the screen of a cathode-ray tube.

Tract : The portion of a moving storage medium which is accessible to a given reading device in a computer.

Tracking : It refers to maintaining a prescribed relationship between an electrical parameter of one electronic circuit or device and the same or a different parameter of a second device, suitably arranged, when both are subjected to the same stimulus. It also refers to maintaining a radar or radio beam set on a target while its position is determined.

Transducer or Sensor : Any device which changes a non electrical parameter, e.g., sound, pressure or light, into electrical signals or a *vice-versa*. The variations in the electrical signal parameter are a function of the input parameter. Transducers are either active or passive. An active transducer is one which introduces gain, in other words it derives energy from a source that is independent of the input signal energy.

Transductor or Saturable Reactor : A device which is made of a megnetic core carrying several windings. The state of megnetic flux density in the core is controlled by a fixed alternating current in one of the windings.

Transfer : It means to transmit or copy information from one device in a computer to another.

Transfer Characteristic : The relation between the degree of illumination of a television camera tube and the corresponding output current under given conditions.

Transfer Current : It refers to the current applied to the starter electrode of a glow-discharge tube in order to initiate a glow discharge acress the main gap.

It also refers to the current at one electrode of gas-filled tube that causes gas breakdown to take place at a different electrode.

Transfer Parameter or Mutual Parameter : The tangent at any point on a given transfer characteristic of any network, amplifying curcuit, transducer, or device.

Transformer : An equipment that has not moving parts and which transforms electrical energy at one alternating voltage into electrical energy at another alternating voltage without change of frequency. It depends for its action upon mutual induction. It consists essentially of two electric circuits coupled together magnetically. The usual construction is of two coils wound on magnetic core. One of these circuits is called the primary.

Transient : A phenomenon, like camped oscillations or a voltage or current surge, which takes place in an electrical system following a sudden change in the dynamic conditions of the systems and that is usually relatively short-lived.

Transistor : A semiconductor device in which the current flowing between two specified electrodes is modulted by the voltage or current applied to one or more specified electrodes.

It is a suitably prepared semiconductor crystal of germanium or silicon. It has replaced the filament valves in wire-less sets and no heater is required.

Modern transistors are of two main classes; bipolar devices, which depends on the folow of both minority and majority carriers through the device and unipolar transistors in which the current is carried by majority carriers only.

Voltage amplification is possible employing the emitter as the input terminal. This method of operation is called common-base connection; the characteristics are shown in Fig. The term 'transistor' is an acronym from transfer resistor and device from this original application of the device since that transfer charactertistic indicates Vomt against Iih and the associated transfer parameter is the resistance (transistance).

Transistors have now replaced thermionic valves as the general-purpose active electronic device, except for some very specialized uses. They are small, robust, require small supply voltages, are relatively easy to manufacture, cheap, and ideally suited for most applications:*n-p-n* transistors are most generally employed and the semiconductor material is usually silicon.

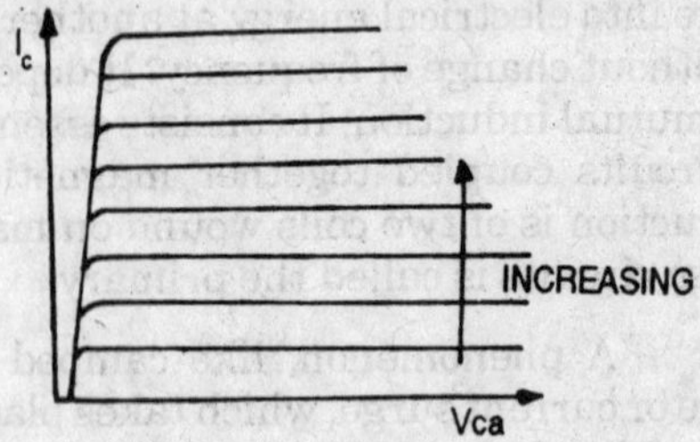

Common-base output characteristics of a junction transistor.

Transistor Parameter : A transistor is a nonlinear device having behaviour which is difficult to represent exactly by a set of mathematical equations.

The transistor is represented by an equivalent circuit with two input terminals and two output terminals (Fig.). This is a quadripole network. The input and output volt-

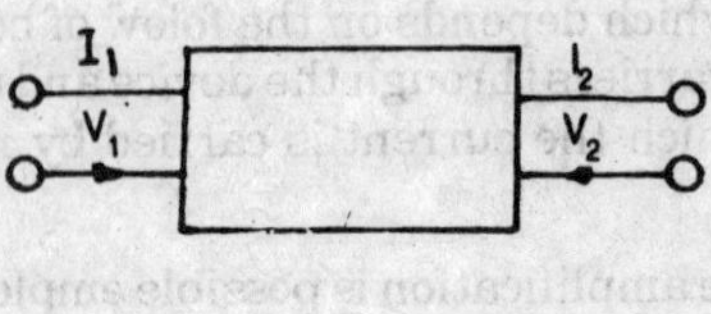

Quadripole network

age and currents can be expressed in general matrix form by two simulatanous equtions.

$[A]=[p][B]$

A and *B* are voltage current and *p* is the parameter of the transistor. The common materix parameters are referred as *h*,*y* and *g* parameters.

The h or hybrid parameters are the most commonly used for bipolar junction transistors; they are so called because the dimensions are mixed and a true matrix representation is not possible.

Other parameters used are z parameters, which are impedance parameters and are the inverse of y parameters; g parameters are the inverse of h parameters.

If the circuit is for use with small signals, the small-signal parameters are written hfe, etc; if for use with large signals, capitals are used for the subscripts: HFE. etc.

Transistor-Transistor Ligic (TTL) : A family of high-speed integrated logic circuits in which the input is through a multi emitter transistor; usually the output stage is push-pull. Diode-transistor ligic operates in a similar manner but the input is through a number of diodes.

The biggest limitation in speed is caused by the delay time due to hole storage in the saturated output transistor. The speed may be imprived by adding a Schottky diode with low forward diode voltage across the base-collector junction of the transistor. This circuit is called Schottky TTL.

TTL is the most widely used type of intergrated logic circuit for high-speed applications and tends to be regarded as a standard against which all other logic circuits are judges.

Transition : A sudden change in the energy state of an atom between two of its energy levels.

Transit Time : The time required for a charge carrier to travel directly from one designated point or region in an electronic device to another, under specified operating conditions.

Translator : It refers to a device which changes information from one form to another.

It also refers to a televisioin transmitter which is peripheral to the main transmitter and is used to extend the service area of the main transmitter.

Transmission : The act of sending information in the form of electrical signals from one designated location to another using a wire, waveguide, transmission line, or radio channel and using any circuits devices, or other equipment that may be necessary.

Transmission Level : It means the ratio of the electical power at any point in a telecomminication system to the power at some arbitray reference point.

Transmission Line or Power Line : An electric line, like an open wire, that is used to convey electrical power from a power station or substation to other stations or sub stations.

Transmission Loss : The reduction in power between any two points 1 and 2 in a telecommunication system given by the ratio *P2/P1*, where *P1* and *P2* are the powers at points 1 and 2 respectively and 2 is further away from the source of power than 1.

In any system where gain is introduced to the signal, the transmission gain is given by the ratio of the powers *P2/P1*.

Transmission Primary : Any of the set of three primary colours employed to give the chrominance signals in a colour-television system. The colours chosen are red, green, and blue.

Transmit-Receive Switch (TR switch) : A switch employed in a radar system having a common aerial for both transmission and reception. The TR switch automatically de couples the receiver from the aerial during the transmitting period. It is mostly used in conjunction with an anti-transmit receive switch (ATR switch) that automatically decouples the transmitter from the aerial during the receiving period; extra protection is thus provided for the transmitter from return echos received by the aerial.

Transmitted-Carrier Transmission : A telecommunication system which employs amplitude modulation of a carrier wave in which the carrier is transmitted.

Transmitter : The device, circuit or equipment used in a telecommunication system to transmit an electrical signal to the receiving part of the system.

Transponder : A combined transmitter receiver system which automatically transmits a signal when a predetermined trigger is received by it. The trigger, which is ofter in the form of a pulse, is called the interrogating signal.

Transposed Tranmission Line : A transmission line having the conductors of which it is composed are interchanged in position at regular intervals of distance.

Tramsverse Wave : A wave having the displacement at each point along the wave ina direction prependicular to the diretion of propagation.

Travelling Wave : An electromagnetic wave which traverls along and guided by a transmission line.

Travelling-Wve-Tube (TWT) or Show Wave Tube : A microwave tube having an electron beam which interacts continuously with a radio-frequency (r.f) electromagnetic field in order to produce amplification or in some type of tube, oscillation at microwave frequencies.

Trickle Change : It is a small continuous charge applied to a secondary battery in ordfer to maintain it in a fully charged conditions during storage.

Trigger : Any stimulus which starts operation of an electronic circuit or device.

Trimmer or Trimming Capacitor : A relatively small variable capacitor employed in parallel with a large fixed capacitor in order to adjust the total capacitance of the combination over a limited range of values.

Triode : A triode is a tube containing three electrodes (i) the filament or cathode (ii)) anode and (iii) grid. The grid is a mesh of fine wires placed between the cathode and the anode plate. The symbol of triode is shown below

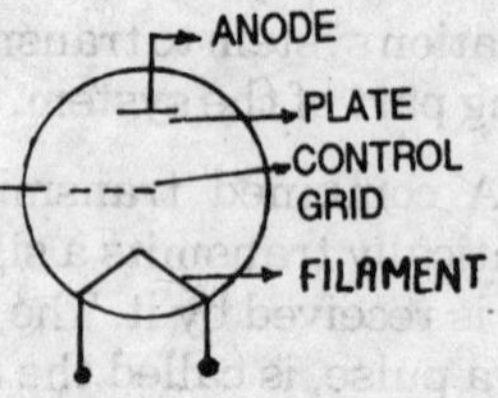

Symbol of TRIODE

Triode-Hexode : A multiple (or multi electrode) thermionic value which has both a triode and a hexode within the same envelope

Triplen Harmonic : A harmonic which has a frequency of 3nf when n is an integer and f the frequency of the fundamental. A frequency multiplier in which the output is the firrst triplen harmonic (n=1) of the input frequency is a tripler.

Trunk Feeder or Interconnecting Feeder; Interconnector; Trunk Trunk Main : A transmission line that is used to interconnect two electric power stations or two electric power distribution networks.

Truth Table : A table used in formal logic which lists the truth or falsity of the outcome when a logical operator, such as 'and' or 'or', is applied to combinations of logical statements.

Tuned Amplifier : An amplifier that is designed to operate only with a very narrow range of input frequencies. Atypical frequency response curve is shown in Fig.a. The response may be modified to give an essentially flat response to the frequencies in the pass band by using stagger tuning (Fig.b).

Tuned Circuit : A resonant circuit having some form of tunning so that the natural resonant frequency of the circuit may be varied. The resonance condition for forced oscillations can be changed. A tuned circuit is employed

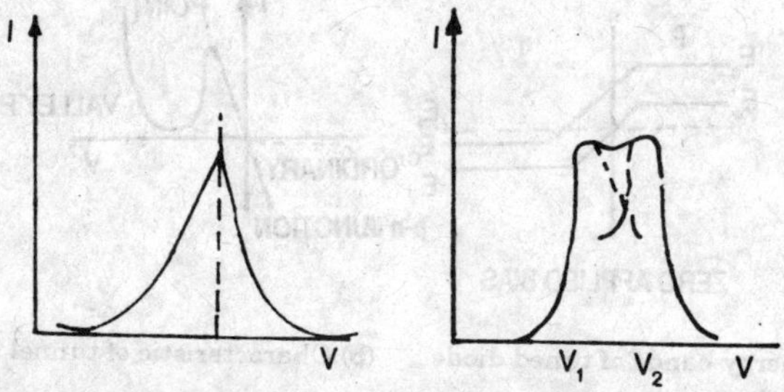

(a) Frequency response of a tuned amplifier (b) Response of a stagger tuned amplifier

to select the frequency or resonance of a variety of frequency-selective devices, like audiofrequency amplifiers.

Tuner : (1) It refers to a device, like a variable capacitor or inductor, which is employed to alter the resonant frequency of a tuned circuit.

(2) It also refers to the first stage of a radio of television receiver that is used to select a particular broadcast channel.

Tungsten Symbol : W. A. heavy metal having atomic number 74, which has an extremely high melting point and is extensively used to form lamp filaments.

Tuned Diode : A *p-n* junction diode having extremely high diping levels on each side of the junction, in other words it is doped into degeneracy. The energy bands are shown in Fig.a for zero applied bias. At such high doping levels tunnelling of electrons across the junction can take place in the forward direction. The tunnel effect takes place for small values of reverse bias but the negative-resistance portion of the characteristic disapperas. The current flowing in the reverse direction is larger than in the forward direction (Fig.d).

Tunnel Effect : It refers to the crossing of a potential barrier by a partial which does not have sufficient kinitic energy to surmount it. The effect if explained by wave mechanics.

Turns Radio or Transformer Ratio : Symbol: v. The m

ratio of the number of turns, *n2* active in the secondary

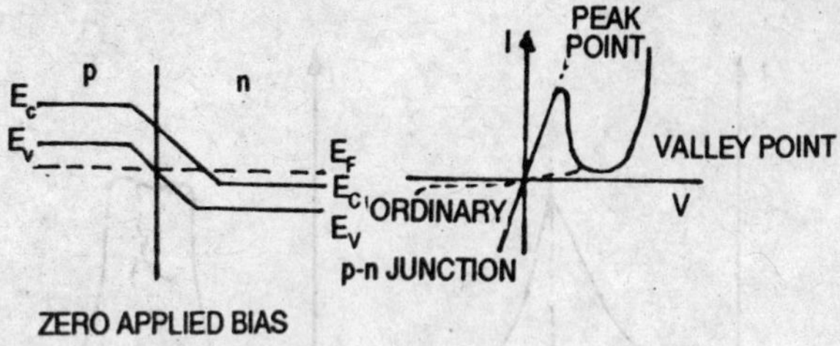

(a) Energy bands of tuned diode (b) Characteristic of tunnel diode

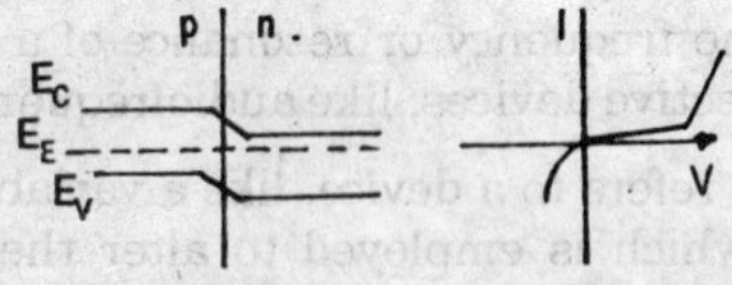

(c) Energy bands of backward diode (d) Characteristic of backward diode

circuit of a transformer to the number of turns *n1* in the primary winding.

Turret Tuner : A tuning device employed in a television or radio receiver. It has a set of resonant circuits each tuned to the frequency of one of the separate broadcast channels.

Tweeter : A physically small loudspeker which reproduces sounds of relatively high frequency.

Two-Way Communication : A telecommunication system by which signals may be transmitted between two points in both directions.

Two-Wire Circuit : A circuit which is made of two conductors insulated from each other and which gives simultaneously a two-way communication channel in the same frequency bandbetween two points in a telecommunication system/

U

Ultrasonic Communication : Underwater communication at ultrasonic frequencies by mans of suitably modified sonar.

Ultrasonic Delay Line : An acoustic delay line which employs acoustic waves of ultrasonic frequencies as the delay element.

Ultrasonic or Supersonics : The study and application of sound frequencies that lie above the limits of audibility of the human ear, (frequencies above 20 kilohertz).

Ultrasonic waves may be detected with the help of hot-wire microphone. The only electrical receiver that is sensitive to ultrasonic wave is the piezoelectric quartz crystal.

The main applications are degassing of melts during the manufacture of electronic components and circuits, cleaning of electronic components, echo sounding and underwater communication, picture transmission using an ultrasonic light valve, and in medicine the location of a brain tumour or a foetus.

Ultraviolet Radiation : Electromagnetic radiation having frequency between light and X-rays on the electromagnetic spectrum.

Unabsorbed Field Strength : The field strength of a radiowave which may exist at the receiving point in the absence of any absorption between the transmitting and receiving aerials.

Undercurrent Release : A device which operates when the current in a circuit falls below a predetermined value.

Undershoot : An initial transient response to a change in an input signal which precedes the desired response and is in the opposite sense.

Undervoltage Release : A device which operates when the voltage in a circuit falls below a predetermined value. A voltage that causes the release to operate is an under voltage.

Underwater Sound Projector : A piezoelectric loud-speaker which is employed to produce sound in water.

Undistored Wave : A periodically varying quantity which is made of sinusoidal components of which none is present at one point which is not present at all points and in which both the attenuation and velocity of propagation are the same for all the sinusoidal components.

Unifilar Suspension : A form of construction of an instrument in which the movable part is suspended by a single thread, wire, or strip.

Uniform line : A transmission line having electrical properties which are substantially indentical along its length.

Uniform Waveguide : A waveguide which has constant electrical and physical characteristics along its axial length.

Unijunction Transistor : Syn. double-base diode. A bipolar transistor that has three terminals and one junction (see diagram).

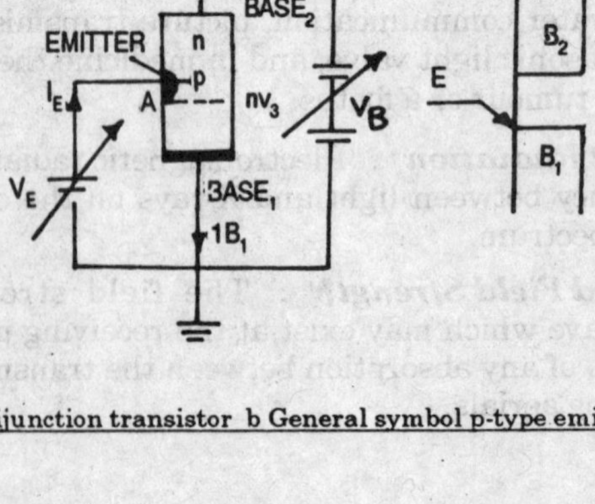

a Unijunction transistor b General symbol p-type emitter

The device has a bar of lightly doped semiconductor, mostly n-type, with an opposite-polarity region of highly doped (low-resistivity) material located near the centre of the bar.

The most common application of the unjunction transistor is in relaxation oscillator circuits.

Unipolar Transistor : A transistor in which current flows due to the movement of majority carriers only.

Unit : A precisely specified quantity, like the ampere or second, in terms of which the magnitude of other physical quantities of the same kind can be written as

physical quantity=numberical value unit

current=n amperes

charge=n amperes per second=n coulombs

In a mechanical system the units can be defined in terms of certain units of the three fundamental quantities of mass, length, and time; other units are derived by multiplying and/or dividing sets of these fundamental units.

Unit-Step Function : An electrical signal which is zero for all times before a predetermined instant and unity for times following that instant.

Universal Motor : An electric motor which is employed with a direct-current power supply or an alternating-current supply.

Universal Shunt : A galvanometer shunt which is tapped so that it can pass designated fractions (0.1,0.01, etc.) of the main current.

V

Vacancy : A site in a crystal lattice which is not occupied by an atomic nucleus.

Vacuum Evaporation : A technique mostly employed during the manufacture of electronic circuits and components in order to give a coating or thin film of one solid material on the clean surface of another, in particular a metal on a semiconductor. The solid to be deposited in heated in a vacuum in the presence of the cool substrate material.

Vacuum Switch : A switch having the contacts contained in a substantially evacuated container in order to minimize spark formation.

Vacuum Tube : It is a device that can control the passage of current through a circuit, much as a valve controls the passage of fluids in a hydraulic system. Such vacuum tubes act as switches and permit only on and off operative.

Valence Band : The band of energies in a solid in which the electrons cannot move freely but are bound to individual atoms.

Valence Electrons : Electrons which occupy the lowest energy levels of an atom and are involved in chemical and physical changes.

Valency : The number of hydrogen (or equivalent) atoms with which an atom will combine or that it will displace. The two most important types of valency are electrovalency and covalency.

Valve or Electron Tube : An active device having two or more electrodes which are enclosed in an envelope, usually of glass, one of the electrodes acting as a primary source of electrons.

Valve Rectifier : A rectifier in which a value, mostly a diode, acts as the rectifying element.

Van de Graaff Accelerator : An accelerator in which the high-voltage terminal of a Van de Graaff genertor acts as a charged-particle source.

Van de Graaff Generator : An electrostatic voltage generator which may produce potentials of millions of volts. An electric charge from an external source is applied to a continuous insulated belt at points. A as shown in the diagram.

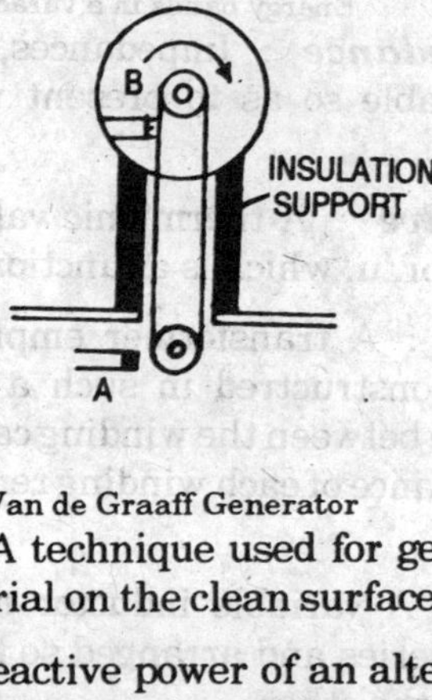

Van de Graaff Generator

Vapour Plating : A technique used for getting a thin film of one solid material on the clean surface of another solid.

Var : The unit of reactive power of an alternating current. It is indentical to the watt, being the product of volts times amperes.

Varactor : A semi conductor diode operated with reverse bias in such a way that it behaves as a voltage-dependent capacitor. The depletion layer at the junction acts as the dielectric and n and p-regions form the plates.

Varactor Tuning : Capacitive tuning used in receivers, like television receivers, in which the variable capacitance element is provided by a varactor.

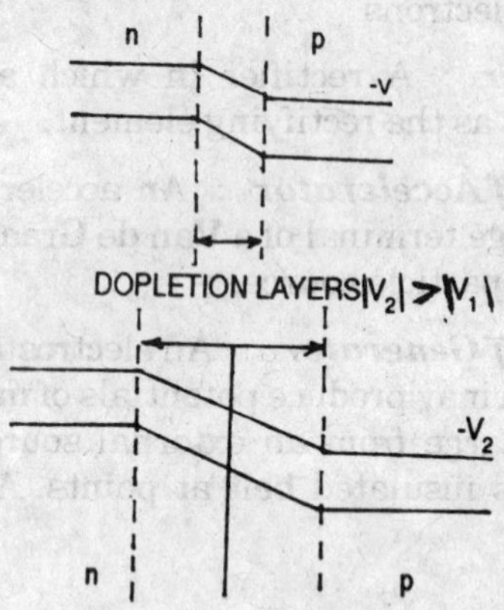

Energy bands in a varactor.

Variable Impedance : Impedances, like capacitors, which are adjustable so as to present variable values of the impedance.

Variable μ Valve : A thermionic valve having the amplification factor, u, which is a function of the grid voltage.

Variocoupler : A transformer employed in radio circuits which is constructred in such a way that the mutual inductance between the winding can be changed while the self-inductance of each winding remains subsantially constant.

Variometer : A variable inductor made of two coils connected in series and arranged so that one coil can rotate relative to the other.

Varistor : A resistor which has a markedly nonohmic characteristic.

V Band : A band of microwave frequencies ranging from 46.0 to 56.0 gigahertz

Vector : A straight line used by engineers to represent a force or velocity an a paper. A vector is considered to name both magnitude and direction so it is sometimes called a directed line.

Vector Diagram : It is a collection of vectors that show the inter-relationship between certain parameters such as force and velocity in a mechanical system or for an electrical circuit between voltage and current. Vector diagrams are usually drawn on a set a axes whose intersection is referred to a zero.

Vector Potential : It is defined as the magnetic vector whose curl at any point gives the value of the magnetic field at that point.

Velocity Microphone : A microphone in which the electrical output is a function of the particle velocity of the detected sound waves.

Velocity Modulation : A process that introduces a radio-frequency (r.f.) component into an electron stream and thereby modulates the velocities of the electrons in the beam.

Vertical Recording : It is a process of recording a gramophone record or magnetic disk in which the groove modulations are orthogonal to both the surface of the recording medium and the motion of the cutter.

Vestigial-Sideband Transmission : It refers to form of suppressed -carrier transmission in which one sideband and the corresponding vestigial sideband are transmitted.

V-Groove Technique : An etching technique employed to give a very precise edge in a silicon crystal. The technique is used to produce mesa layers or during the manufacture of V/MOS circuits.

Vibrator : A device which generates an alternating current by periodically interrupting or reversing a continuous steady current from a direct-current source.

Video Frequency : It means the frequency of any component of the output signal from a television camera. Video frequencies are within the range 10 hertz to two megahertz.

Video Mapping : A method employed in the display of a radar system in which a chart or map of the area covered by the radar is superimposed electronically on the radar display.

Videophone : A telephone receiver which can receive and display visible images simultaneously with the telephone signals.

Videotape : A form of magnetic tape that is suitable for use with a television camera.

Many television programmes are recorded on to videotape prior to transmission rather than being broadcast live.

Vidicon : A low-electron velocity photoconductive camera tube which is widely used in closed-circuit television and as an outside broadcast camera since it is smaller, simpler, and cheaper than the image orthicon.

The photosensitive target area of the vidicon has a transparent conducting film placed on the inner surface of the thin glass faceplate.

Virtual Cathode : It means the surface, located in a space-charge region between the electrodes of a thermionic valve, at which the electric potential is a mathematical minimum and the potential gradient is zero.

Visibility Factor : The ratio of the minimum descernible signal (mds) of a television or radar receiver which may be detected by ideal instruments to the mds that can be detected by a human operator.

Visual Display Unit (VDU) : A display instrument employed with a computer that displays information in the form of character and line drawings on the screen of a cathode-ray tube.

V/MOS : MOS circuits or transistors that are fabricated using the V-groove technique.

Voice-Frequency Telephony : A form of telephony in which the electric signals are transmitted as audiofrequency signals of substantially the same frequencies as the voice frequencies producing them.

Volt : Symbol V. It is defined as the potential difference between two points on a conductor when the current flowing is one ampere and the power dissipated between the points is one watt. Generally volts are measured by comparison with a Weston standard cell using a potentiometer.

Voltage : Symbol: V, unit: volt. IT means the potential difference between two points in a circuit or device.

Voltage between Lines or Line Voltage, Voltage between Phases : The voltage between the two lines of a single-phase electrical power system or between any two lines of a symmetrical three-phase system.

Voltage Doubler : An arrangement of two rectifiers which gives an output voltage amplitude twice that of a single rectifier. In a typical circuit using diode rectifiers (see diagram) each rectifier separately rectifies alternate half cycles of the input alternating voltage and the two outputs are then summoned

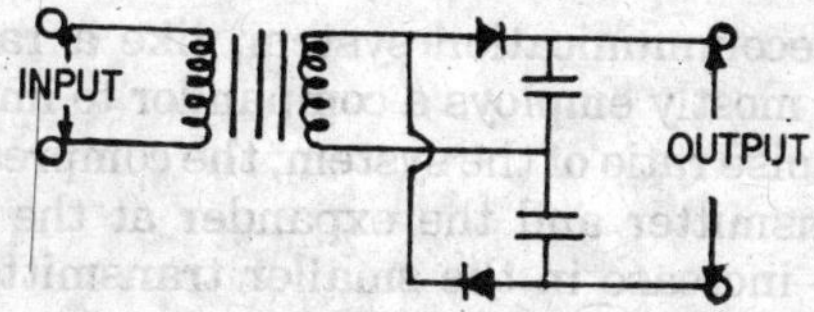

Voltage doubler

Voltage Drop : The voltage between any tow specific points of an electrical conductor.

Voltage Generator : Any source of electrical power which is employed to supply voltage to a circuit or device.

Voltage Jump : An abrupt undesired change or discontinuity in the operating voltage of any electronic circuit or device.

Voltage Level : The ratio of the voltage at a point in a transmission system to the voltage of an arbitrary reference point.

Voltage Multiplier : An arrangement of rectifiers which gives an output voltage amplitude which is an integral multiple of that of a single rectifier, i.e. of the peak value of applied alternating voltage.

Voltage Stabilizer : A circuit or device which gives an output voltage substantially constant and independent of variations either in the input voltage or in the load current.

Volt-Ampere : Symbol: VA. The unit of apparent power, defined as the product of the root-mean-square values of voltage and current in an alternating-current circuit.

Voltmeter : An instrument which measures voltage. Voltmeters in common use include d.c. instruments digital voltmeters and cathode-ray oscilloscopes.

Volt Per Meter : The SI unit of electric field strength.

Volume : The magnitude of the complex audiofrequency signals in an audiofrequency transmission system.

Volume Compressor : An instrument which automatically reduces the range of amplitude variations of an audiofrequency signal in a transmission system.

A telecommunication system, like a radio-telephone system, mostly employs a compandor to improve the signal-to-noise ratio of the system; the compressor is used at the transmitter and the expander at the receiver. The relative increase in the smaller transmitted signals reduces the effect of noise on these signals.

W

Wafer: : It is defined as a large single crystal of semiconductor material, usually silicon, that is used as the substrate during the manufacture of a number of chips.

Wagner Earth Connection : This is regarded as a method of connection used with an alternating current bridge circuit that minimizes the admittance to earth of the bridge. The earth connection is formed by means of the

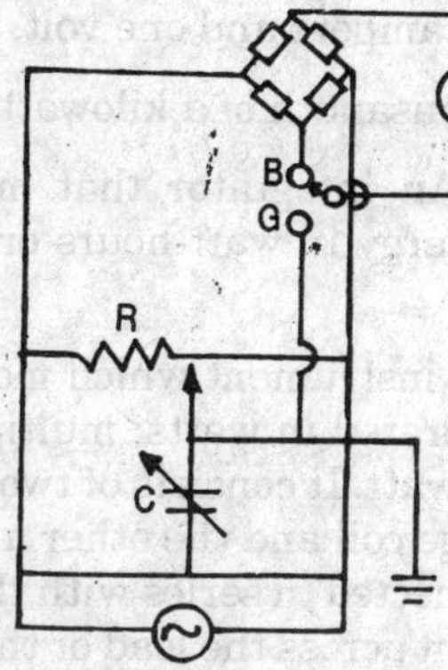

Wagner Earth Connection

centre tapping of a three-terminal adjustment resistor, R, in parallel with the a.c. supply.

Walkie-Talkie : It is defined as a light portable radio set designed to be carried and capable of allowing two-way communication by any person while in motion.

Walk-Out : This is regarded as a phenomenon observed in semiconductor devices that are subjected to repeated avalanche breakdown. It is progressive increase in the value of the avalanche-breakdown voltage of the device and due to the injection of hot electrons into the surface passivating oxide layer.

Wall Effect : It is an effect which is due to the inside wall of a container and contributes to the behavious of the system contained by it.

Warble : It is defined as the periodic variation of a frequency between two limits. The frequency variation is generally small compared to the nominal frequency and is repeated several times per second.

Warm-Up : This is the period after a particularly electronic device, circuit, or apparatus has been switched on during which it reaches a state of thermal equilibrium with its surroundings.

Watt : Its symbol is W. The SI unit of power. It is defined as the power resulting when one joule of energy is dissipated in one second. In an electric circuit one watt is given by the product of one ampere and one volt.

Watt-Hour : One thousandth of a kilowatt-hour.

Watt-Hour Meter : An integrator that measures and records electrical energy in watt-hours or more usually in kilowatt-hours.

Wattmeter : It is an instrument which measures electrical power and is calibrated in watts, multiples of a watt, or submultiples of a watt. It consists of two coils of wire, one is known as voltage coil and the other a current coil. The current coil is connected in series with the circuit and the voltage coil is shunt across the load or the other section to be measured. A pointer indicates directly on metre scale the power in the circuit.

Wave : It is defined as a periodic disturbance, either continuous or transient, which travels through a medium or through space and in which the displacement from a mean value is a function of time or position or both.

The wavelength is the distance between two displacements of the same phase along the direction of propogation. If *n* is the velocity of the wave and *I* its wavelength, then the frequency of vibration, *n* is given by

$$v=v/\lambda$$

The frequency is the reciprocal of the period, *T*, of the wave. The frequency (or wavelength) of electromagnetic radiation is commonly used to describe particular regions of the electromagnetic spectrum, such as the visible or radio regions.

Wave Analyser : It is a device such as a spectrum analyser or frequency analyser, that resolves a complex waveform into its fundamental and harmonic components.

Waveguide : It is defined as a transmission line which is made of a suitably shaped hollow conductor, which may be filled with a dielectric material, and this is used to guide ultra high-frequency electromagnetic waves propagated along its length.

For any given transmission mode there is a lower limit to the frequency that may be propagated through the

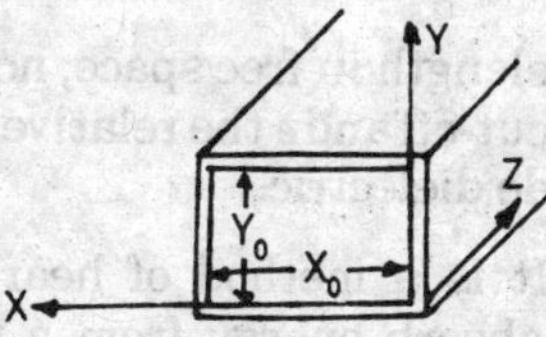

(a) Rectangular waveguide

waveguide. This cut-off frequency results when the complex propagation coefficient becomes real. The electromagnetic wave is then at tenuated exponentially and soon becomes substantially zero. The waveguide therefore acts effectively as a high-pass filter. In an ideal (lossless) waveguide, above the cut-off frequency, the propagation coefficient is purely imaginary and no attenuation occurs.

The most common shapes of waveguide are rectangular (Fig.a) and cylindrical; the most common dielectric is air. A cylindrical waveguide is called a wave duct; one which contains a solid rod of dielectric is a uniconductor waveguide. If a wide range of frequencies is to be transmitted a ridged waveguide (Fig.b) may be used.

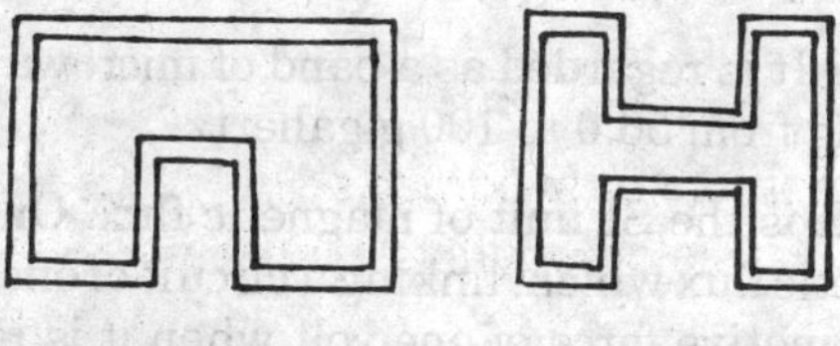

(b) Cross Section of a ridged waveguides.

Waveguides are mostly employed for the transmission of electromagnetic waves. Electromagnetic energy may also be stored by using a section of waveguide short-circuited at each end of form a cavity resonator. A section of coaxial transmission line may also be used. When the length of the section of waveguide or transmission line is h, resonance take place when

h=n(λg'e)

where *n* is an integer and

λg=λ[E-(λ/λc)2]1/2

n is the wavelength in free space, nc the wavelength of the waveguide cut-off and *e* the relative dielectric constant of the waveguide dielectric.

Wave Heating : It is a method of hearing a material by allowing it to absorb energy from a travelling wave of electromagnetic radiation.

Wavemeter : This is an apparatus that measures the frequency or wavelength of a radiowave. It is made of a capacitively tuned circuit together with a current-detecting instrument. The variable capacitor is calibrated directly to read frequency or wavelength. A current maximum is obtained within the resonant frequency of the tuned circuit corresponds to the frequency of the received radiowave.

Wavetrain : It is defined as a succession of waves, particularly a small group of waves of limited duration.

Wavetrap : It is a tuned circuit, which is included in a radio receiver in order to reduce interference from unwanted radiowaves of the particular frequency to which the circuit is tuned.

W Band : It is regarded as a band of microwave frequencies ranging from 56.0 to 100 gigahertz.

Weber : It is the SI unit of magnetic flux. One weber is the magnetic flux which, linking a circuit of one turn, gives an electromotive force of one volt when it is reduced to zero at a uniform rate in one second.

Welding : It is regarded as the process of joining together two pieces of the same metal using either no extra metal or extra metal of the same material.

Wenner Winding : It is a method of forming a wire-wound resistor in order to reduce the reactance.

Weston Standard Cell : It is syn. to cadmium cell. A cell which contains a substantially constant terminal voltage and is used as a reference standard for elecromotive force.

Wheatstone Bridge : It is a four-arm bridge used for measuring resistance. Each arm contains a resistance (see diagram), with the unknown and reference resistances, R_1 and R_2, being connected at a point. At balance, when a null response is obtained from the indicating instrument, we have

$R_1 / R_2 = R_3 / R_4$

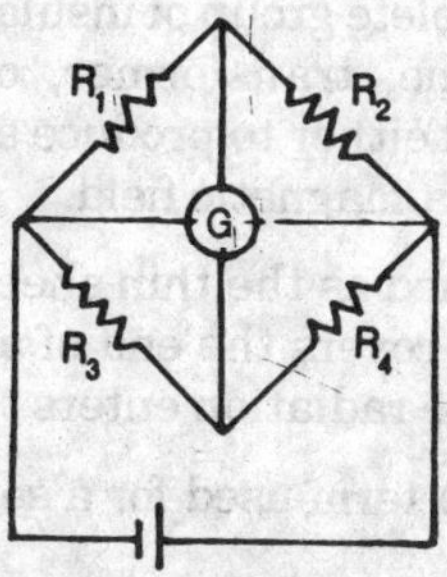

Wheatstone Bridge

The two arms R_3 and R_4 are known as the ratio arms and may take the form of a wire of uniform resistance, which is tapped by using a sliding contact. R_3 and R_4 are proportional to the lengths of wire l_1 and l_2 on each side of the contact, we have

$R_1/R_2 = I_1/I_2$.

White Compression : Compression applied to the components of a television video signal which correspond to the light areas of the picture.

Wideband : It is defined as a frequency band that extends over a relatively large range.

It also denotes an electronic device or circuit, such as an amplifier, that operates satisfactorily over a large range of input signal frequencies.

Wien Bridge : It is a four-arm bridge that is used for the measurement either of capacitance or frequency. A typical network is shown in the diagram. At balance, when a null response is obtained from the indicating instrument, we have

$C_x C_3 = (Ro/Ra) - R_3/R_x)$

where w is the angular frequency. For the measurement of a frequency f, it is convenient to make

Wien Effect : It is the increase in the conductivity of an electrolyte when subjected to a high voltage gradient.

Wilson Effect : The effect refers to electrical polarization of an insulator when it is moved through a region containing a magnetic flux.

Winding : It is complete group of insulated conductors in an electrical machine, transformer, or other equipment which is designed either to produce a magnetic field or to be acted upon by a magnetic field.

Window : It is defined as the thin sheet of material, generally, mica, which covers the end of a Geiger counter and through which the radiation enters the instrument.

Wire Gauge : It is a term used for a set of standard diameters of wire.

Wiring Diagram : It is a diagram of electronic equipment showing the interconnections between assemblies and subassemblies.

Wobbulator : It is signal generator in which an automatic periodic variation is applied to the output frequency so that it varies over a predetermined range of values. It may be used to investigate the frequency response of an electronic circuit or device or as a test instrument for a tuned circuit.

Wollaston Wire : It is a very fine platinum wire used for electroscope wires, microfuses, and hot-wire instruments.

Woofer : It is a relatively large loudspeaker that is used to reproduce sounds of relatively low frequency in a high-fidelity sound reproduction system.

Work : It is a force acting through a distance. When a mechanical force overcomes a friction and makes the body move, work is being done. In electric circuits when the *emf* overcomes the resistance and current appears in the circuit work is being done.

Work Function : Its symbol is Φ ; unit: electonvolt. It is designated as the difference in energy between the Fermi level of a solid and the energy of free space outside the solid, i.e. the vacuum level.

Wow : It refers to low-frequency (up to about 10 hertz) audible periodic variations in the pitch of the sound output in a highfidelity sound-resproduction system. Wow is an undesired form of frequency modulation and is generally caused by a non-uniform rate of reproduction of the original sound, such as by non uniform rotation of a gramophone turnable.

Write : To enter information into a storage element of a memory of storage tube.

X

X-Axis : It is defined as the horizontal axis on the screen of a cathode-ray tube.

X Band : It is designated as a band of microwave frequencies ranging from 5.2 to 10.9 gigahertz

Xerography : It is regarded as a photographic copying process that uses electrical effects to form the image. An image of the document to be copied is reproduced on a uniformly charged plate, mostly coated with selenium: different areas of the plate are discharged to different extents by an intensity-modulated beam of ultaviolet radiation to leave a charge pattern corresponding to be brightness information of the original document.

X-Rays : The term is used for electromagnetic radiation of frequencies between that of ultraviolet radiation and gamma rays, and sometimes higher. They are obtained when matter is bombarded with sufficiently energetic electrons. X-rays can be obtained as a result of electron transitions from higher to lower energy levels within an atom. X-rays are also produced during the rapid decelerations of electrons as they approach the nucleus. These X-rays, have a relatively wide frequency range and are known as *continuous x-rays*. X-rays of relatively low energy are termed soft X-rays; those at the high energy portion of the frequency spectrum are hard X-rays.

X-rays can be reflected, reflected, and polarized and also show interference and diffraction. They interact with matter to produce relatively high energy electrons the mechanism is the same as in the photoelectric effect. These electrons are mostly of sufficiently high energy to ionize a gas or to produce secondary X-rays from the matter

X-rays are widely used in radiography to investigate materials that are opaque to visible light but relatively transparent to X-rays. Radiography is used to detect flaws in structures and for diagnostic examinations. Sufficiently high intensities of X-rays are used in radiotherapy for the therapeutic destruction of diseased tissue. X-rays, are also used in X-rays crystallography to investigate and determine crystal structures by using the crystal as a three-dimensional diffraction grating.

X-Ray Tube : It is an electron tube that produces X-rays. Early X-ray tubes were gas-filled tubes but only electron beams of relatively low energy could be obtained due to disruptive discharge occurring through the gas. Modern X-ray tubes are invariably hard tubes developed from the Coolidge type of tube; they are more stable than the earlier gas-filled sort.

X-Y Plotter : It is a graphical instrument that produces a chart showing the relationship between two varying signals. One of the signals causes the pen to move in the direction of the x-axis and the other independently causes it to move in the direction of the y-axis.

Y

Yagi Aerial : It is defined as a sharply directional aerial array from which most aerials used for television and radioastronomy have been developed.

Y-Axis : It is designated as the vertical axis on the screen of a cathode ray tube.

YIG : This abbre, for yttrium-iron garnet. A ferrite that is widely used for microwave applications. The magnetic properties are altered by the amount of trace elements in the material. The most common trace elements are calcium, vanadium and bismuth. Calcium-germanium YIGs may be easily grown on a substrate.

Yoke : It is regarded as a piece of ferromagnetic material which is employed to connect permanently two or more magnetic cores and complete a magnetic circuit without surrounding it by a winding of any kind.

Z

Zener Breakdown : It is regarded as a type of breakdown observed in a reverse-biased *p-n* junction that has very high doping concentration of both sides of the junction.

Zener Diode : This is a special form of *PN* Diode designed to permit recurrent applications of breakdown voltage to the diode without damaging the junction.

Zero Level : An arbitrary reference level employed in telecommunication systems to compare the relative intensities of transmitted signals or of noise.

Zinc : Its symbol: Zn. It is a metal, atomic number 30, which is used as an electrode in some electrolytic cells.

Zirconium : Its symbol is Zr. It is a metal, atomic number 40, that can be used as a getter in hard vacuum electron tubes.

Zone Refiding : It is defused as a method of redistributing impurities within a solid material, like a semiconductor, by melting parts of the material and causing the molten zones to move along sample.